This report contains the collective views of an international group of experts and does not necessarily represent the decisions or the stated policy of the United Nations Environment Programme, the International Labour Organisation, or the World Health Organization.

Environmental Health Criteria 113

FULLY HALOGENATED CHLOROFLUOROCARBONS

Published under the joint sponsorship of the United Nations Environment Programme, the International Labour Organisation, and the World Health Organization

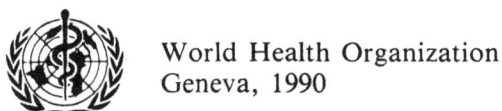

World Health Organization
Geneva, 1990

The **International Programme on Chemical Safety (IPCS)** is a joint venture of the United Nations Environment Programme, the International Labour Organisation, and the World Health Organization. The main objective of the IPCS is to carry out and disseminate evaluations of the effects of chemicals on human health and the quality of the environment. Supporting activities include the development of epidemiological, experimental laboratory, and risk-assessment methods that could produce internationally comparable results, and the development of manpower in the field of toxicology. Other activities carried out by the IPCS include the development of know-how for coping with chemical accidents, coordination of laboratory testing and epidemiological studies, and promotion of research on the mechanisms of the biological action of chemicals.

WHO Library Cataloguing in Publication Data

Fully halogenated chlorofluorocarbons.

(Environmental health criteria ; 113)

1. Freons - adverse effects   2. Freons - toxicity   I. Series

ISBN 92 4 157113 6   (NLM Classification: QV 633)
ISSN 0250-863X

©World Health Organization 1990

Publications of the World Health Organization enjoy copyright protection in accordance with the provisions of Protocol 2 of the Universal Copyright Convention. For rights of reproduction or translation of WHO publications, in part or *in toto,* application should be made to the Office of Publications, World Health Organization, Geneva, Switzerland. The World Health Organization welcomes such applications.

The designations employed and the presentation of the material in this publication do not imply the expression of any opinion whatsoever on the part of the Secretariat of the World Health Organization concerning the legal status of any country, territory, city, or area or of its authorities, or concerning the delimitation of its frontiers or boundaries.

The mention of specific companies or of certain manufacturers' products does not imply that they are endorsed or recommended by the World Health Organization in preference to others of a similar nature that are not mentioned. Errors and omissions excepted, the names of proprietary products are distinguished by initial capital letters.

Printed in Finland
90/8487 — Vammala — 5000

CONTENTS

ENVIRONMENTAL HEALTH CRITERIA FOR FULLY HALOGENATED CHLOROFLUOROCARBONS

1. SUMMARY 13

    1.1 Identity, physical and chemical properties, analytical methods 13
    1.2 Sources of human and environmental exposure 14
    1.3 Environmental transport, distribution, and transformation 15
    1.4 Environmental levels and human exposure 15
    1.5 Kinetics and metabolism 16
    1.6 Effects on the environment 17
    1.7 Effects on experimental animals and *in vitro* systems 18
    1.8 Effects on humans 19
    1.9 Evaluation of human health risks 21

2. IDENTITY, PHYSICAL AND CHEMICAL PROPERTIES, ANALYTICAL METHODS 22

    2.1 Identity 22
        2.1.1 Technical product 22
    2.2 Physical and chemical properties 25
    2.3 Conversion factors 26
    2.4 Analytical methods 26

3. SOURCES OF HUMAN AND ENVIRONMENTAL EXPOSURE 30

    3.1 Natural occurrence 30
    3.2 Man-made sources 30
        3.2.1 Production levels 30
        3.2.2 Manufacturing processes 33
        3.2.3 Loss during disposal of wastes 34
        3.2.4 Release from transport, storage, and accidents 34
            3.2.4.1 Transport and storage 34
            3.2.4.2 Accidents 35
    3.3 Use patterns 35
        3.3.1 Major uses 35

| | | 3.3.2 | Release from use: controlled or uncontrolled | 36 |

4. ENVIRONMENTAL TRANSPORT, DISTRIBUTION, AND TRANSFORMATION — 39

4.1 Transport between media — 39
4.2 Environmental transformation and degradation processes — 39
    4.2.1 Oxidation — 39
    4.2.2 Hydrolysis — 39
    4.2.3 Photolysis — 40
        4.2.3.1 Photochemistry — 40
        4.2.3.2 Environmental transformation — 40
    4.2.4 Biodegradation — 41
4.3 Interaction with other physical, chemical, or biological factors — 41
4.4 Bioconcentration and bioaccumulation — 42

5. ENVIRONMENTAL LEVELS AND HUMAN EXPOSURE — 43

5.1 Environmental levels — 43
    5.1.1 Air — 43
    5.1.2 Water — 46
    5.1.3 Food and other edible products — 46
5.2 Occupational exposure — 46

6. ECOLOGICAL EFFECTS OF STRATOSPHERIC OZONE DEPLETION — 47

6.1 Introduction — 47
6.2 Terrestrial plants — 49
6.3 Aquatic organisms — 51
6.4 Research needs — 53

7. KINETICS AND METABOLISM — 56

7.1 Absorption — 56
7.2 Distribution — 58
7.3 Metabolic transformation — 59
7.4 Elimination and excretion in expired air, faeces, and urine — 60

7.5	Retention and turnover	60
7.6	Reaction with body components	62

8. EFFECTS ON EXPERIMENTAL ANIMALS AND *IN VITRO* TEST SYSTEMS ... 63

- 8.1 Single exposures ... 63
  - 8.1.1 Acute inhalation toxicity ... 63
  - 8.1.2 Acute oral toxicity ... 63
- 8.2 Short-term exposures ... 63
  - 8.2.1 Inhalation exposure ... 65
  - 8.2.2 Oral toxicity ... 72
  - 8.2.3 Dermal toxicity ... 73
- 8.3 Skin and eye irritation; sensitization ... 74
- 8.4 Long-term exposures ... 74
  - 8.4.1 Inhalation toxicity ... 74
  - 8.4.2 Oral toxicity ... 75
- 8.5 Reproduction and developmental toxicity ... 77
  - 8.5.1 Reproduction ... 77
  - 8.5.2 Developmental toxicity ... 77
- 8.6 Mutagenicity and related end-points ... 79
- 8.7 Carcinogenicity ... 79
- 8.8 Special studies - cardiopulmonary effects ... 83
  - 8.8.1 Cardiac sensitization in response to exogenous adrenaline-induced arrhythmia ... 83
  - 8.8.2 Cardiac sensitization and asphyxia-induced arrhythmia ... 85
  - 8.8.3 Arrhythmia not associated with asphyxia or adrenaline ... 86
- 8.9 Mechanisms of toxicity - mode of action ... 87

9. EFFECTS ON HUMANS ... 89

- 9.1 Controlled studies with volunteers ... 89
- 9.2 Occupational exposure ... 92
- 9.3 Non-occupational exposures ... 94
- 9.4 Health effects associated with stratospheric ozone depletion ... 95
  - 9.4.1 Skin cancer effects ... 95
  - 9.4.2 Immunotoxic effects ... 98
  - 9.4.3 Ocular effects ... 99
  - 9.4.4 Effects on vitamin D synthesis ... 99

	9.4.5	Exacerbation of photochemical smog formation and effects	99

10. EVALUATION OF HUMAN HEALTH RISKS AND EFFECTS ON THE ENVIRONMENT — 100

    10.1 Evaluation of human health risks — 100
        10.1.1 Direct health effects resulting from exposure to fully halogenated chlorofluorocarbons — 100
        10.1.2 Health effects expected from reduction of stratospheric ozone by chlorofluorocarbons — 101
    10.2 Effects on the environment — 102
    10.3 Conclusions — 103

11. RECOMMENDATIONS — 104

REFERENCES — 106

RESUME — 129

EVALUATION DES RISQUES POUR LA SANTE HUMAINE ET EFFETS SUR L'ENVIRONNEMENT — 139

RECOMMANDATIONS — 144

RESUMEN — 147

EVALUACION DE LOS RIESGOS PARA LA SALUD HUMANA Y DE LOS EFFECTOS EN EL MEDIO AMBIENTE — 157

RECOMENDACIONES — 162

WHO TASK GROUP ON ENVIRONMENTAL HEALTH CRITERIA FOR FULLY HALOGENATED CHLOROFLUOROCARBONS

Members

Dr B. Gilbert, Company for the Development of Technology Transfer, Cidade Universitaria, Campinas, Brazil (*Rapporteur*)

Professor H.A. Greim, Institute of Toxicology and Biochemistry, Association for Radiation and Environmental Research, Neuherberg, Federal Republic of Germany (*Chairman*)

Dr L. Hinkova, Toxicologist, Institute of Hygiene and Occupational Health, Sofia, Bulgaria

Dr Y. Lessard, Laboratory of Medical Physiology, Faculty of Medicine, University of Rennes, France

Dr M. Morita, Department of Legal Medicine, Sapporo Medical College, Sapporo, Japan

Dr G. Neumeier, Federal Office for the Environment, Berlin, Federal Republic of Germany

Professor M. Noweir, Occupational Health Research Centre, Higher Institute of Public Health, University of Alexandria, Alexandria, Egypt

Dr J. Sokal, Department of Toxicity Evaluation, Institute of Occupational Medicine, Lodz, Poland

Professor J.C. Van der Leun, Institute of Dermatology, State University Hospital of Utrecht, Utrecht, Netherlands

Dr K. Victorin, National Institute of Environmental Medicine, Department of Environmental Hygiene, Stockholm, Sweden

Dr W.D. Wagner, National Institute of Occupational Safety and Health, Cincinnati, Ohio, USA

Members (contd)

Dr R.C. Worrest, Stratospheric Ozone Research Program, Office of Environmental Processes and Effects Research, US Environmental Protection Agency, Washington, D.C., USA

Observers

Dr D. Mayer, Toxicology Department, Hoechst AG, Frankfurt am Main, Federal Republic of Germany

Dr H. Trochimowicz, E.I. Du Pont de Nemours, Haskell Laboratory for Toxicology and Industrial Medicine, Newark, Delaware, USA

Representatives of Host Country

Dr U. Schlottmann, Federal Ministry for the Environment, Nature Conservation and Nuclear Safety, Bonn, Federal Republic of Germany[b]

Dr V. Quarg, Federal Ministry for the Environment, Nature Conservation and Nuclear Safety, Bonn, Federal Republic of Germany[b]

Secretariat

Professor F. Valić, IPCS Consultant, World Health Organization, Geneva, Switzerland (*Responsible Officer and Secretary*)[a]

Dr S. Lutkenhoff, Office of Health and Environmental Assessment, US Environmental Protection Agency, Cincinnati, Ohio, USA

Dr G. Quélennec, Division of Vector Biology and Control, World Health Organization, Geneva, Switzerland

---

[a] Vice-rector, University of Zagreb, Zagreb, Yugoslavia.
[b] Present for part of the meeting only.

NOTE TO READERS OF THE CRITERIA MONOGRAPHS

Every effort has been made to present information in the criteria monographs as accurately as possible without unduly delaying their publication. In the interest of all users of the environmental health criteria monographs, readers are kindly requested to communicate any errors that may have occurred to the Manager of the International Programme on Chemical Safety, World Health Organization, Geneva, Switzerland, in order that they may be included in corrigenda, which will appear in subsequent volumes.

* * *

A detailed data profile and a legal file can be obtained from the International Register of Potentially Toxic Chemicals, Palais des Nations, 1211 Geneva 10, Switzerland (Telephone No. 7988400 or 7985850).

ENVIRONMENTAL HEALTH CRITERIA FOR FULLY HALOGENATED CHLOROFLUOROCARBONS

A WHO Task Group on Environmental Health Criteria for Fully Halogenated Chlorofluorocarbons met at the Institute of Toxicology and Biochemistry, Neuherberg, Federal Republic of Germany, from 21 to 25 November 1988. Professor H.A. Greim opened the meeting on behalf of the host institute. Dr U. Schlottmann spoke on behalf of the Federal Government, which sponsored the meeting. Professor F. Valić welcomed the members on behalf of the three cooperating organizations of the IPCS (UNEP/ILO/WHO). The Task Group reviewed and revised the draft criteria monograph and made an evaluation of the risks for human health and the environment from exposure to fully halogenated chlorofluorocarbons.

The drafts of this monograph were prepared by the Office of Health and Environmental Assessment, US Environmental Protection Agency, under the direction of Dr J. STARA and Dr S. LUTKENHOFF. The chapter on the ecological effects of stratospheric ozone depletion was prepared by Dr R.C. WORREST and the section on the health effects associated with stratospheric ozone depletion by Dr L. GRANT, both of the US Environmental Protection Agency. Professor F. Valić and Dr P.G. Jenkins (IPCS) were responsible for the overall scientific content and editing, respectively.

ABBREVIATIONS

ADI	Acceptable daily intake
ADP	Adenosine diphosphate
bw	Body weight
CFC	Chlorofluorocarbon
EC	Electron capture
ECG	Electrocardiogram
EEG	Electroencephalogram
FEV	Forced expiratory volume
FI	Flame ionization
GC	Gas chromatography
HCFH-22	Chlorodifluoromethane ($CHClF_2$)
LDH	Lactate dehydrogenase
LOEL	Lowest-observed-effect level
MS	Mass spectrometry
NMR	Nuclear magnetic resonance
NOEL	No-observed-effect level
ppb	Parts per billion
ppm	Parts per million
ppt	Parts per trillion
SGOT	Serum glutamic oxaloacetic transaminase
SGPT	Serum glutamic pyruvic transaminase

TWA	Time-weighted average
UV	Ultraviolet
v/v	Volume per volume
w/v	Weight per volume

1. SUMMARY

1.1 Identity, physical and chemical properties, analytical methods

This monograph concerns only those chlorofluorocarbons (CFCs) that are derived from the complete substitution of the hydrogen atoms in methane and ethane with both fluorine and chlorine atoms. Many of these compounds are of commercial significance and some of them are known to contribute to ozone depletion. Compounds considered in this report include: trichlorofluoromethane (CFC-11), dichlorodifluoromethane (CFC-12), chlorotrifluoromethane (CFC-13), 1,2-difluoro-1,1,2,2-tetrachloroethane (CFC-112), 1,1-difluoro-1,2,2,2-tetrachloroethane (CFC-112a), 1,1,2-trichloro-1,2,2-trifluoroethane (CFC-113), 1,1,1-trichloro-2,2,2-trifluoroethane (CFC-113a), 1,2-dichloro-1,1,2,2-tetrafluoroethane (CFC-114), 1,1-dichloro-1,2,2,2-tetrafluoroethane (CFC-114a), and 1-chloro-1,1,2,2,2-pentafluoroethane (CFC-115). Compounds not containing chlorine have not been considered. Those compounds containing hydrogen will be reviewed in a subsequent report.

Commercial chlorofluorocarbons rank among the highest purity organic chemicals available. They are usually characterized by high vapour pressure and density and low viscosity, surface tension, refractive index, and solubility in water. The degree of fluorine substitution greatly affects the physical properties and, in general, as fluorine substitution increases, the vapour pressure increases, and the boiling point, density, and solubility in water decrease.

The chlorofluorocarbons reviewed in this monograph are reasonably stable chemically and, in the absence of metal catalysts, exhibit low rates of hydrolysis. They are highly resistant to attack by conventional oxidizing agents at temperatures below 200 °C. In general, chlorofluorocarbons show a high degree of thermal stability and are extremely resistant to almost all chemical reagents. However, they will interact violently with chemically reactive metals.

Summary

Several analytical methods are available for the determination of chlorofluorocarbons in various media. These include spectrophotometry, gas chromatography with several quantification methods, and mass spectrometry. The majority of methods utilize gas chromatography with various detection techniques, and detection limits are often of the order of 1 part per trillion (ppt). Methods for sample collection have been modified to achieve greater selectivity and sensitivity.

1.2 Sources of human and environmental exposure

The chlorofluorocarbons discussed in this monograph are not known to occur naturally in the environment, but practically all chlorofluorocarbons, except those used as chemical intermediates, are released into the environment. The estimated world production of the important potentially ozone-depleting chlorofluorocarbons (CFC-11, CFC-12, CFC-113) in 1985 was at least a million tonnes. Manufacture is not limited to major industrial nations; it occurs in at least 16 countries. With the implementation of the Montreal Protocol, the present growth trend in the production of these chlorofluorocarbons will probably be reversed.

The most important method for manufacturing the major chlorofluorocarbons is the catalytic displacement of chlorine from chlorocarbons with fluorine by reaction with anhydrous hydrogen fluoride. Most release to the environment occurs during the disposal of waste refrigerant-containing equipment, and not during manufacture, storage, or handling. The release of propellant chlorofluorocarbons has decreased as a result of legislative restrictions on their use in many countries, and the release of blowing agents is small. Because of the high vapour pressure of these compounds at ambient temperatures, almost all of the amount released into the environment eventually accumulates in the atmosphere. The estimated total annual release of about one million tonnes consisted in 1985 largely of CFC-11 and CFC-12, and the cumulative release of these chlorofluorocarbons from 1931 to 1985 was about 13.5 million tonnes.

The approximate world use pattern of chlorofluorocarbons in 1985 was as follows: refrigerants, 15%; foam-blowing agents, 35%; aerosol propellants, 31%; miscellaneous, 7%, and unallocated, 12%. In the USA, the aerosol propellant use was much lower because of restrictions.

1.3 Environmental transport, distribution, and transformation

The commercial chlorofluorocarbons are persistent in the environment because of their chemical stability. The average residence times in the atmosphere are estimated to be 65, 110, 400, 90, 180, and 380 years for CFC-11, CFC-12, CFC-13, CFC-113, CFC-114, and CFC-115, respectively. These long residence times will ensure diffusion into the stratosphere where, via photochemically-produced chlorine atoms, the chlorofluorocarbons will react with the ozone layer. Additionally, these compounds will contribute to the greenhouse effect.

1.4 Environmental levels and human exposure

The global distribution of chlorofluorocarbons has been reported by several investigators. Recent measurements of latitudinal variations of chlorofluorocarbon concentrations indicate little difference in CFC-11 and CFC-12 concentrations between the northern and southern hemispheres. Also there is no significant variation with altitude up to 6 km above the Earth's surface. The measured concentrations of chlorofluorocarbons in urban/suburban air are higher than those in rural/remote areas because of contributions from local sources of emission.

Atmospheric levels of CFC-11 and CFC-12 increased steadily up to 1985, when combined levels for these two compounds in the USA were 9120 ng/m^3 in urban/suburban areas and 2720 ng/m^3 in rural/remote areas for both compounds. From these data, human inhalation intake has been estimated at 182 and 54 mg/day in these two types of areas.

The mean surface ocean concentrations of CFC-11 and CFC-12, reported from three mutually distant locations, were of the order of 0.2 ng/litre. However, 0.62 ng CFC-11 per litre was measured in the Greenland Sea in 1982 and values of up to 0.54 ng/litre have been measured in Japanese coastal waters. The highest value for CFC-12 reported was 0.33 ng/litre in these same coastal waters.

Summary

Much higher levels have been measured in fresh water in Lake Ontario where 249 mg CFC-11 per litre and 572 ng CFC-12 per litre have been recorded. Chlorofluorocarbons have not been detected in drinking-water, but have been found in snow and rain water in Alaska, in Lake Ontario, and in the Niagara river. CFC-11 has been detected at levels of 0.1-5 µg/kg (ppb) (dry weight basis) in various organs of fish and molluscs. However, the presence of chlorofluorocarbons in processed food has not been documented.

1.5 Kinetics and metabolism

Chlorofluorocarbons may enter the human organism by inhalation, ingestion, or dermal contact. Inhalation is the most common and important route of entry, and exhalation is the most significant route of elimination from the body. Controlled studies with volunteer subjects and experimental animals have provided substantial data from exposures to a number of the chlorofluorocarbons. These data indicate that chlorofluorocarbons:

- can be absorbed across the alveolar membrane, gastrointestinal tract, or the skin;
- are absorbed rapidly into the blood, following inhalation;
- are absorbed into the blood at a decreasing rate as blood concentration increases;
- once in the blood, are absorbed by various tissues;
- will reach a stable blood level if exposure is sufficiently long, indicating an equilibrium between the air containing the chlorofluorocarbons and the blood;
- are still absorbed by body tissue, after the initial blood level stabilization, and continue to enter the body.

Studies with animals indicate that chlorofluorocarbons are rapidly absorbed after inhalation and are distributed by blood into practically all tissues of the body. The highest concentrations are usually found in fatty or lipid-containing tissues. However, chlorofluorocarbons are also found in organs with a good blood supply, e.g., heart, lung, kidney, muscle.

Results from animal and human metabolic studies have demonstrated the resistance of chlorofluorocarbons to breakdown or metabolic transformation in biological systems. These results suggest that chlorofluorocarbons, in general, are metabolized to a very small degree, if at all, following exposure.

Regardless of the route of entry, chlorofluorocarbons are eliminated almost exclusively through the respiratory tract via exhaled air. No significant recovery of chlorofluorocarbons or their metabolites has been reported in studies attempting to identify metabolic transformation products via elimination in urine or faeces.

1.6 Effects on the environment

Certain chlorofluorocarbons, including CFC-11, 12, 113, 114, and 115, are extremely stable under conditions found in the lower atmosphere. It is not until these gases migrate into the high-energy radiation environment of the upper stratosphere that photolytic processes split the chlorine off from the chlorofluorocarbons. These chlorine radicals catalytically destroy ozone. Stratospheric ozone absorbs solar ultra-violet radiation (UV-B: 280-320 nm wavelength) allowing only reduced UV-B radiation to penetrate to the surface of the earth.

Experimental evidence suggests that increased UV-B irradiation at the Earth's surface, resulting from ozone depletion, would have deleterious effects on both terrestrial and aquatic biota. Despite uncertainties resulting from the complexities of field experiments, the data currently available suggest that crop yields and forest productivity are vulnerable to increased levels of solar UV-B radiation. Existing data also suggest that increased UV-B radiation will modify the distribution and abundance of plants, and change ecosystem structure.

Various studies of marine ecosystems have demonstrated that UV-B radiation causes damage to fish larvae and juveniles, shrimp larvae, crab larvae, copepods, and plants essential to the marine food web. These damaging effects include decreased fecundity, growth, and survival. Experimental evidence suggests that even small increases in

Summary

ambient UV-B exposure could result in significant ecosystem changes.

1.7 Effects on experimental animals and *in vitro* systems

The acute inhalation toxicity of chlorofluorocarbons has been extensively studied. The chlorofluorocarbons considered in this monograph show low acute inhalation toxicity. The symptomatology of acute intoxication involves CNS effects, secondary effects on the cardiovascular system, and irritation of the respiratory tract. The limited information available on the acute oral toxicity of chlorofluorocarbons indicates low toxicity. When applied dermally in high doses, CFC-112, CFC-112a, and CFC-113 cause various degrees of irritation but no other significant effects.

Short-term inhalation studies have been reported for CFC-11, CFC-12, CFC-112, CFC-113, CFC-114, and CFC-115. The results showed low toxicity, and the effects observed were related mainly to the CNS, respiratory tract, and the liver. Oral toxicity studies have confirmed the low toxicity.

In a long-term inhalation study, rats were exposed to CFC-113 at 0.2, 1, or 2% (15.3, 76.6, or 183 g/m^3) 6 h per day, 5 days/week for up to 2 years. No histopathological effects or changes in clinical laboratory values were observed. The only finding considered by the authors to be treatment-related was a reduction in body weight gain in the groups exposed to the two highest doses.

The available information indicates that the fully halogenated chlorofluorocarbons evaluated in this monograph have little or no mutagenic or carcinogenic potential. Negative results have been obtained *in vitro* using bacteria and mammalian cells with or without metabolic activation and in the dominant lethal test.

Long-term carcinogenicity studies (by oral and inhalation routes) with CFC-11 and CFC-12 in rats and mice showed negative results. Although a tumorogenic response in the nasal cavity was observed in rats upon inhalation of CFC-113, this response was considered equivocal. The tumours were of various morphologies and the incidences were not dose-related.

Of the eight chlorofluorocarbons reviewed in this document, developmental toxicity studies have been reported in the available scientific literature for CFC-11, CFC-12, and CFC-113. No evidence of embryotoxicity, fetotoxicity, or teratogenicity has been documented for any of these three chlorofluorocarbons.

1.8 Effects on humans

Controlled studies of volunteers using CFC-11 and CFC-12 revealed no observable effects on clinical haematology and chemistry, EEG, or neurological parameters.

At high concentrations, subjects experienced a tingling sensation, humming in the ears, and apprehension. EEG changes were noted as well as slurred speech and decreased performance in psychological tests. An exposure to an 11%[a] (545 g/m^3) concentration of CFC-12 for 11 min caused a significant degree of cardiac arrythmia, followed by a decrease in consciousness with amnesia after 10 min.

Following exposure to CFC-12 at a concentration of 1% (50 g/m^3) for 150 min, a 7% decrease in psychomotor test scores was noted, but no effects were observed at 0.1% (5 g/m^3).

In a study in which 10 subjects were exposed to CFC-11, CFC-12, CFC-114, two mixtures of CFC-11 and CFC-12, and a mixture of CFC-12 and CFC-114 (breathing concentrations between 16 and 150 g/m^3) for 15, 45, or 60 seconds, significant acute reduction of ventilatory lung capacity (FEF50, FEF25) was reported in each case, as well as bradycardia and increased variability in heart rate and atrioventricular block.

Psychomotor performance was evaluated using CFC-113 at concentrations of 0.15% (12 g/m^3), 0.25% (19 g/m^3), 0.35% (27 g/m^3), or 0.45% (35 g/m^3) for 165 min. There was no effect at the lowest concentration, but there was difficulty in mental concentration and some decrease in test scores beginning at 0.35% (27 g/m^3).

---

[a] Throughout this monograph, percentages of chlorofluorocarbons in air are expressed as the volume of chlorofluorocarbon divided by the volume of air.

Limited studies indicate that individuals with a prior history of skin reaction to deodorant sprays containing CFC-11 or CFC-12 may become sensitized to dermal applications of certain chlorofluorocarbons. The tracheal mucociliary function in five non-smokers was not impaired by exposure to CFC-11.

Two studies suggest that normal occupational exposure to CFC-113 does not pose a serious health hazard. No adverse effects occurred at occupational levels as high as 0.47% (36.7 g/m^3), with an average level of 0.07% (5.4 g/m^3).

Although chlorofluorocarbons have been used for over 50 years, only one cohort study (539 exposed workers) is available. No increase in total deaths or tumour deaths was observed.

Significant acute reduction in the ventilatory lung capacity of hairdressers using chlorofluorocarbon-containing hairsprays was observed in several studies. Cases of neurological effects attributed to occupational exposure to chlorofluorocarbons have been reported. One case of neuropathy in a laundry worker, exposed to tetrachloroethene and to undetermined levels of CFC-113 for 6 years, has been described.

Non-occupational exposure and accidental or abusive inhalation of aerosols have also been documented, the main symptoms being CNS depression and cardiovascular reactions. Cardiac arrythmia, possibly aggravated by elevated levels of catecholamines due to stress or by moderate hypercapnia, is suggested as the cause of these adverse responses, which may lead to death.

Increased UV-B radiation is expected to lead to predominantly adverse effects on human health, but the state of knowledge varies greatly from one effect to another. It is virtually undisputed that the incidence of non-melanoma skin cancers will increase. Projections based on recent data indicate that the incidence of non-melanoma skin cancers will increase by 3% for every 1% depletion of ozone. On this basis, an ozone depletion by 5% would lead, after several decades, to about 240 000 additional new cases of non-melanoma skin cancer per year, worldwide.

UV-B radiation appears also to play a role in the formation of the more dangerous cutaneous melanomas. However, there is insufficient knowledge to determine accurate dose-response relationships.

The immune system is influenced by UV-B radiation in various ways. Although the knowledge available is insufficient to predict the consequences of ozone depletion for human health, increased incidence of some infectious diseases might be one of the consequences.

The most important effect for the human eye would be an increase in the incidence of cataracts, a permanent clouding of the eye lens which leads, even at current levels of UV-B radiation, to impaired vision and blindness in many people.

Increased UV-B radiation would be expected to increase photochemical smog, and this would aggravate the related health problems in urban and industrialized areas.

1.9 Evaluation of human health risks

The most important direct effects on humans from exposure to chlorofluorocarbons are caused by the excessive concentrations resulting from industrial accidents or poor occupational practices and from misuse or abuse of the chemicals when used as solvents or as propellant gases. Release of chlorofluorocarbons into the global environment during use, disposal of wastes, transport, and storage are an increasing concern because of the potential impact such uncontrolled releases may have on the future health of mankind, mainly through the depletion of stratospheric ozone.

2. IDENTITY, PHYSICAL AND CHEMICAL PROPERTIES, ANALYTICAL METHODS

2.1 Identity

The chlorofluorocarbons (CFCs) considered in this monograph are compounds derived by the complete substitution of the hydrogen atoms in methane and ethane with both fluorine and chlorine atoms. Chlorofluorocarbons containing hydrogen (designated HCFC) will be reviewed in a subsequent report. The chemical formulae, relative molecular masses, common names, common synonyms, and CAS Registry numbers of some of the chlorofluorocarbons reviewed (CFCs 11, 12, 13, 112, 112a, 113, 113a, 114, 114a, 115) are given in Table 1.

Chlorofluorocarbons are marketed under many different trade names, e.g., Algcon, Algofrene, Arcton, Eskimon, Flugene, Forane, Freon, Frigen, Genetron, Isceon, Osotron, Khladon. The individual chemical substances are characterized by code numbers, as defined in DIN 89 62, which are very widely adopted and uniformly used.

2.1.1 Technical product

Commercial chlorofluorocarbons rank among the highest purity organic chemicals sold in the USA (Bower, 1973), the purity of commercial CFC-11 and CFC-12 commonly exceeding 99.9% (Hamilton, 1962). The predominant isomers of the ethane series (CFC-113, CFC-114) are the more symmetrical ones ($CCl_2F.CClF_2$ and $CClF_2.CClF_2$). CFC-113 usually contains no more than a few tenths of 1% of CFC-113a ($CCl_3.CF_3$), while CFC-114 usually contains no more than 7-10% $CCl_2F.CF_3$. Levels of other impurities in the four major CFCs (CFC-11, CFC-12, CFC-113, CFC-114) are: moisture, 10 ppm; residue, a few ppm; acids, much less than 1 ppm; and non-condensibles (i.e., air components) 100-200 ppm in the liquid phase or 0.5-1.0% in the gas phase (Hamilton, 1962).

The commercial chlorofluorocarbons may also be formulated with chemicals other than CFCs, such as actone, ethanol, isopropanol, and methylene chloride. In addition,

Table 1. Identity and physical and chemical properties of commercially significant fully halogenated chlorofluorocarbons[a]

Chemical formula	CCl_3F	CCl_2F_2	$CClF_3$	$CCl_2F \cdot CCl_2F$	$CCl_3 \cdot CClF_2$
Relative molecular mass	137.37	120.92	104.46	203.82	203.82
Common name	trichlorofluoro-methane	dichlorodifluoro-methane	chlorotrifluoro-methane	1,2-difluoro-1,1,2,2-tetrachloroethane	1,1-difluoro-1,2,2,2-tetrachloroethane
CAS registry number	75-69-4	75-71-8	75-72-9	76-12-0	76-11-9
Common synonyms and trade names	CFC-11, F-11, Freon 11, Frigen 11, Arcton 9	CFC-12, F-12, Freon 12, Arcton, Frigen 12, Genetron 12, Halon, Osotron 2	CFC-13, F-13	CFC-112, F-112	CFC-112a, F-112a
Physical state	liquid at temperatures < 23.7 °C	gas	gas	solid	solid
Colour	colourless	colourless	colourless	white	
Odour	faint ethereal	nearly odourless	ethereal	slightly camphor-like	
Melting point (°C)	-111	-158	-181	26	40.6
Boiling point (°C)	23.82	-29.79	-81.4	92.8	91.5
Flashpoint[b]	NF	NF	NF	NF	NF
Density of saturated vapour at boiling point (g/litre)	5.86	6.33	7.01	7.02	
Solubility in water (25 °C) (wt %)	0.11	0.028	0.009	0.012 (saturation pressure)	
Conversion factor (ppm(v/v) → mg/m³) (20 °C)	5.71	5.03	4.34	8.47	8.47

Table 1 (contd).

Chemical formula	$CCl_2F.CClF_2$	$CCl_3.CF_3$	$CClF_2.CClF_2$	$CCl_2F.CF_3$	$CClF_2.CF_3$
Relative molecular mass	187.38	187.38	170.92	170.92	154.47
Common name	1,1,2-trichloro-1,2,2-trifluoroethane	1,1,1-trichloro-2,2,2-trifluoroethane	1,2-dichloro-1,1,2,2-tetrafluoroethane	1,1-dichloro-1,2,2,2-tetrafluoroethane	1-chloro-1,1,2,2,2-pentafluoroethane
CAS registry number	76-13-1	354-58-5	76-14-2	374-07-2	76-15-3
Common synonyms and trade names	CFC-113, F-113 Freon 113	CFC-113a	CFC-114, F-114	CFC-114a, F-114a	CFC-115, F-115 Freon 115
Physical state	liquid	liquid	gas	gas	gas
Colour	colourless		colourless		colourless
Odour	nearly odourless		nearly odourless		
Melting point (°C)	-35	14.2	-94	-94	-106
Boiling point (°C)	47.57	45.8	3.77	3.6	-39.1
Flashpoint[b]	NF	NF	NF	NF	NF
Density of saturated vapour at boiling point (g/litre)	7.38		7.83		8.37
Solubility in water (25 °C) (wt %)	0.011		0.009		0.006
Conversion factor (ppm(v/v) → mg/m³) (20 °C)	7.79	7.79	7.11	7.11	6.42

[a] From: Du Pont (1980b); Smart (1980); Hawley (1981); and Windholz (1983).
[b] NF: non-flammable.

nitromethane or other stabilizers are sometimes added to alcohol-based aerosols (0.3% by weight) (Du Pont, 1980a).

2.2 Physical and chemical properties

Chlorofluorocarbons are usually characterized by high vapour pressure (low boiling point) and density and low viscosity, surface tension, refractive index, and solubility in water. The common physical and chemical properties of the commercially significant chlorofluorocarbons are given in Table 1.

The degree of fluorine substitution greatly affects the physical properties of chlorofluorocarbons. In general, as the number of fluorine atoms replacing chlorine increases, the vapour pressure also increases, but the boiling point, density, and solubility in water decrease. For example, in the chlorofluoroethane series, vapour pressure increases with fluorination in the sequence:

CFC-112 < CFC-113 < CFC-114 < CFC-115 < CFC-116

The solvent power of the chlorofluorocarbons ranges from poor for the highly fluorinated compounds to fairly good for the less fluorinated compounds (Du Pont, 1980b). Being typical non-polar liquids, they exhibit low water solubility.

Apart from their use as chemical intermediates, the chlorofluorocarbons reviewed find applications that reflect their chemical stability rather than chemical reactivity. This chemical stability is a result of the strength of the C-F bond (Bower, 1973).

Although quite inert, chlorofluorocarbons do exhibit some chemical reactivity in some applications. For example, although they exhibit a low rate of hydrolysis compared with other halogenated compounds, the rate of hydrolysis is greatly affected by temperature, pressure, the presence of metals, and the pH of the solution (Du Pont, 1980a,b). Thus CFC-11 is considered unsuitable for water-based products packaged in metal containers since some metals may catalyse the hydrolysis of CFC-11 with liberation of acid. Sanders (1960) has demonstrated a free-radical reaction between CFC-11 and alcohols resulting in dichloromonofluoromethane and small amounts of tetrachlorodifluoroethane. The reaction is inhibited by high concentrations of oxygen and, therefore, it is unlikely that it will occur in nature. In some cases dechlorination by zinc (also by magnesium and aluminium) can occur in the presence of polar solvents:

$$FCl_2C\text{-}CClF_2 \xrightarrow[\text{Alcohol}]{\text{Zn}} FClC=CF_2 + ZnCl_2$$

Chlorofluorocarbons are highly resistant to attack by conventional oxidizing agents at temperatures <200 °C (Downing, 1966; Bower, 1973). In general, they exhibit a high degree of thermal stability, but when pyrolysis occurs in the presence of humidity the products usually include hydrofluoric and hydrochloric acid and, in the presence of either water or oxygen, phosgene.

The photolysis of chlorofluorocarbons is discussed in section 4.2.3.

The carbon-fluorine bonds in chlorofluorocarbon compounds are extremely resistant to almost all chemical reagents. Reduction with hydrogen does not occur until temperatures are >830 °C, and often the C-C bond is also cleaved. Strong reducing agents such as lithium aluminium hydride will not reduce the C-F bond. However, chlorofluorocarbons react violently with alkali and alkaline earth metals, such as sodium, potassium, and barium (Bower, 1973).

2.3 Conversion factors

Conversion factors for the chlorofluorocarbons reviewed in this monograph are given in Table 1.

2.4 Analytical methods

Several analytical procedures used for the determination of chlorofluorocarbons are summarized in Table 2. Methods used include spectrophotometry, gas chromatography with several quantification procedures, and mass spectrometry. However, the majority of methods use gas

Table 2. Analytical methods for the determination of chlorofluorocarbons

Sample type	Sampling method/clean-up	Analytical method	Detection limit v/v	Reference
Air		modified inlet with silicon rubber membrane; mass spectrometry	100 ppb	Collins & Utley (1972)
Air		gas chromatography - electron capture detection	50-100 ppb	Collins et al. (1965)
Air		gas chromatography - electron capture detection	5-10 ppt	Lovelock et al. (1973); Su & Goldberg (1973); Hester et al. (1974)
Air	sorption on cold (< -50 °C) activated carbon	gas chromatography - electron capture detection	1 ppt	Parylczak et al. (1985)
Air	sorption on cold (< -50 °C) Tenax-TA + activated carbon	gas chromatography - electron capture detection	1 ppt	Reineke & Baechmann (1985)
Air	cryogenic trapping in porous glass beads	gas chromatography - electron capture detection	1 ppt	Rudolph & Jebsen (1983)
Air	sorption on cold (liquid N_2) SE-30/glass wool	gas chromatography - electron capture detection	1 ppt	Singh et al. (1983)

Table 2 (contd).

Sample type	Sampling method/clean-up	Analytical method	Detection limit v/v	Reference
Air	sorption on cold (-40 °C) OV-101	gas chromatography - electron capture detection	1 ppt	Makide et al. (1980)
Air	sorption on cold (< -50 °C) activated carbon	gas chromatography - high-resolution mass spectrometry	2.6 ppt	Crescentini et al. (1983)
Air		absorption spectrometry using diode laser		Zasavitskii et al. (1984)
Air	sorption on Tenax GC/ activated carbon	capillary column gas chromatography - electron capture detection		Hanai et al. (1984)
Air (occupational)		spectrophotometry of pyridine complex	7 ppm	Tyras (1981)
Sea water	dynamic purge and trap	gas chromatography - electron capture detection	0.003 ng/litre	Bullister & Weiss (1983)
Blood	head space	gas chromatography - electron capture detection	0.01-0.01 ng/litre	Ramsey & Flanagan (1982)

chromatography with various detection techniques. Methods for sample collection have been developed to achieve greater selectivity and sensitivity.

3. SOURCES OF HUMAN AND ENVIRONMENTAL EXPOSURE

3.1 Natural occurrence

The chlorofluorocarbons discussed in this monograph are not known to occur in nature.

3.2 Man-made sources

Almost all chlorofluorocarbons produced, except for those used as chemical intermediates, are eventually released into the environment, whether during manufacture, handling, use, or disposal. The significance of the release mechanisms discussed below should be evaluated with this in mind.

3.2.1 Production levels

The estimated world production of the three important potentially ozone-depleting chlorofluorocarbons (CFC-11, CFC-12, and CFC-113) in 1985 was approximately one million tonnes, about 30% being in the USA (SRI, 1986). Table 3 indicates some of the major world producers in 1985 (CMA, 1986; CMR, 1986; Rand, 1986; SRI, 1986).

The reported total demand for all chlorofluorocarbons in the USA in 1985 was 458 000 tonnes (CMR, 1986), a 26% increase from the demand figure in 1980 (CMR, 1981). Production figures for CFC-11, CFC-12, and CFC-113 in the USA for 1974-1985 are given in Table 4. In 1984, these three CFCs accounted for 83% of the total chlorofluorocarbons produced in the USA (US ITC, 1985). Based on the 1980 demand and the strong market position in several applications, CMR (1986) projected that the demand for chlorofluorocarbons in the USA would grow to 458 000 tonnes in 1985 and reach 590 000 tonnes by 1990, a positive growth in this 5-year period of 5% per year. However, this was before the Montreal Protocol was signed in September

Table 3. Some of the major world producers of chlorofluorocarbons in 1985[a,b]

Country	Company name
Argentina	Ducilo S.A. (subsidiary of Du Pont de Nemours and Co.)
Australia	Pacific Chemical Industries Pty. Ltd. (subsidiary of Atochem S.A.); Australian Fluorine Chemical Pty. Ltd.
Brazil	Du Pont do Brasil S.A. (subsidiary of Du Pont de Nemours and Co.); Hoechst do Brasil Quimica e Farmacêutica S.A. (subsidiary of Hoechst A.G.)
Canada	Allied Canada, Inc. (subsidiary of Allied Corp.); Du Pont Canada Inc.
France	Atochem S.A.
Germany, Federal Republic of	Hoechst AG (Frigen); Kali-Chemie AG (Kaltron)
Greece	Société des Industries Chimiques du Nord de la Grèce, S.A.
India	Navin Fluorine Industries
Italy	Montefluos S.p.A. (Algofren)
Japan	Asahi Glass Co., Ltd. (Asahiflon); Daikin Kogyo Co., Ltd. (Daiflon); Du Pont Mitsui Fluorochemical Co., Ltd. (Flon) Showa Denko, K.K.
Mexico	Quimoleasicos, S.A. (subsidiary of Allied Corp.); Halocarburos S.A. (subsidiary of E.I. Du Pont de Nemours and Co., Inc.)
Netherlands	Akzochemic B.V.; Du Pont de Nemours (Nederland) B.V. (subsidiary of E.I. Du Pont de Nemours and Co., Inc)
Spain	Ugimica S.A. (subsidiary of Atochem, S.A.); Hoechst Iberica (subsidiary of Hoechst AG); Kali-Chemie S.A. (subsidiary to Kali-Chemie AG)
United Kingdom	Imperial Chemical Industries PLC (Arcton); I.S.C. Chemicals Ltd. (Isecon)
USA	Allied Corp.; E.I. Du Pont de Nemours and Co. Inc.; Essex Chemical Corp.; Kaiser Aluminum and Chemical Corp.; Pennwalt Corp.
Venezuela	Produren (subsidiary to Atochem, S.A.)

[a] From: CMA (1986); CMR (1986); and SRI (1986).
[b] Trade names are given in parentheses, where available.
From: Noble (1972) and Smart (1980).

Table 4. Production of the major chlorofluorocarbons in the USA
in thousands of tonnes[a]

Year	CFC-11	CFC-12	CFC-113[b]
1985	73.9[b,c]	127.9[b,c]	73.2[c]
1984	83.9	152.7	65.9[d]
1982	63.6	117.0	NA
1980	71.7	133.8	NA
1979	75.8	133.3	NA
1978	87.9	148.4	NA
1977	96.4	162.5	>23.1
1976	116.2	178.3	NA
1975	122.3	178.3	NA
1974	154.6	221.1	29.0

[a] From: US ITC (1975-85), unless otherwise specified.
[b] From: Smart (1980), US EPA (1980), and Rand (1986).
[c] It is assumed that consumption was the same as production volume.
[d] Estimated value from the 1985 production data and the assumption that the 1984 production volume was 10% lower (CMR, 1986).
NA = Not available.

1987.[a] The demand for CFC-11, which is used mainly for foam blowing, was largely anticipated to follow the expansion pattern of the construction industry. Demand for fluoropolymers made from CFC-113 (as well as from HCFCs 22 and 142b) is expected to grow at a rate of 10% or more because of electrical and electronic applications. The demand for CFC-113 is also expected to grow because of its use as a solvent in the semi-conductor industry and as a replacement for chlorinated solvents under regulatory pressure (CMR, 1986). Between 1964 and 1974, the production of CFC-11 and CFC-12 increased at 8 and 9% per year respectively. At that time, the hypothesis that certain chlorofluorocarbons that accumulate in the upper atmosphere could deplete the earth's ozone layer had a major impact on the fluorochemical industry (Smart, 1980).

---

[a] The Montreal Protocol on Substances that Deplete the Ozone Layer, signed by 24 countries in September 1987, requires a 20% reduction in use and production of the chlorofluorocarbons 11, 12, 113, 114, and 115 from 1 July 1993 and a further 30% reduction from 1 July 1998. It stipulates a number of stepwise importation bans binding on signatories in order to achieve these reductions (United Nations Environment Programme. Montreal Protocol on Substances that Deplete the Ozone Layer, Final Act, Montreal, 1987).

The US EPA (1978) ruled that most aerosol products containing CFC-11 and CFC-12 propellants could not be manufactured in the USA after 15 December, 1978. As a result, the production of CFC-11 and CFC-12 fell sharply, stabilizing in 1980. However, with the entry into force of the Montreal Protocol, which progressively limited the production of CFCs-11, 12, 113, 114, and 115, the release of all of these chlorofluorocarbons should decline.

3.2.2 Manufacturing processes

The traditional method for manufacturing the fully halogenated chlorofluorocarbons is the catalytic displacement of chlorine from chlorocarbons with fluorine by reaction with anhydrous hydrogen fluoride (Hamilton, 1962; Smart, 1980). Carbon tetrachloride, and hexachloroethane (or tetrachloroethylene plus chlorine) are commonly used starting materials for 1- and 2-carbon chlorofluorocarbons. Carbon tetrachloride is normally used for producing CFC-11, CFC-12, and CFC-113. The reaction can occur in either liquid or vapour phases. The processes use antimony pentafluoride or an equivalent catalyst, in contact with which the chlorocarbon and hydrogen fluoride react. Excess hydrogen fluoride may then be recovered and the chlorofluorocarbon stream is neutralised to remove traces of acid and dried. The chlorofluorocarbons are then separated in a fractionating column and sent to storage. An alternative process for the production of the methane-based chlorofluorocarbons uses the direct reaction of methane with a mixture of chlorine and hydrogen fluoride (Noble, 1972). Other commercially important chlorine-fluorine-substituted hydrocarbons are manufactured by similar processes (Lowenheim & Moran, 1975).

The production processes described above give very high yields. Losses of chlorofluorocarbons are limited to small mechanical leakage, small amounts leaving with the by-product hydrogen chloride, and miscellaneous venting. The total material loss is estimated to be 1% at most (McCarthy, 1973) for the production operations excluding transport and storage. Fuller et al. (1976) assumed a total production loss of 1.5% for the commercially produced chlorofluorocarbons.

3.2.3 Loss during disposal of wastes

The release of chlorofluorocarbons into the environment during their disposal arises mainly from pre-fabricated refrigeration and air-conditioning equipment. Environmental contamination due to chlorofluorocarbon disposal results principally from the following:

- Unreclaimed refrigerants in the cooling systems of scrapped pre-fabricated-type refrigeration and air-conditioning units. Disposal of these old appliances is usually to scrap yards or waste dumps. Efforts are made in some countries to remove chlorofluorocarbon refrigerants before discarding equipment.

- Discarding of vessels containing unused chlorofluorocarbons.

- Time-release of trapped blowing agents in rigid urethane products. This is a minor source of environmental contamination compared with that of scrapped refrigerants.

Waste disposal streams resulting from manufacturing operations are very minor contamination sources compared with scrapped refrigerants.

Because of the high vapour pressure of chlorofluorocarbons at ambient temperature, all releases pass eventually into the atmosphere except in cases where the compounds have been chemically altered.

3.2.4 Release from transport, storage, and accidents

3.2.4.1 Transport and storage

The principal factor required for the transport and storage of the major chlorofluorocarbons is adequate design to meet the elevated pressures. The products are shipped in a wide variety of pressure containers ranging from 23-litre drums to 91-m^3 tank cars.

The containers are fitted with safety valves, rupture discs, and fusible plugs according to US Interstate

Commerce Commission (ICC) specifications; also included are requirements for labelling and leak pressure testing.

Loss of product during transport and storage is relatively minor because of the completely closed system used. Losses are further controlled by monitoring discrepancies, if any, between product billings and receipts. In addition, the high cost of the products provides an incentive to control losses. The total industry-wide loss in transport and storage is <1% of the total quantity produced.

3.2.4.2 Accidents

Data concerning accidental release are not readily available. However, it is probable that quantities released by accident are negligible compared with quantities released by use and disposal.

3.3 Use patterns

3.3.1 Major uses

Chlorofluorocarbons are commercially important because of their unique physical and chemical properties and relatively low physiological activity. They are mainly used as refrigerants, solvents, blowing agents, sterilants, aerosol propellants, and as intermediates for plastics. Table 5 lists the estimated use patterns of chlorofluorocarbons in the USA for the years 1975, 1978, 1981, and 1985. The aerosol propellant market, which consumed half of the total chlorofluorocarbon production in 1975, is currently a minor application because of governmental restrictions.

Estimated use patterns of CFC-11 and CFC-12 in the USA and worldwide (excluding eastern European countries) are given in Tables 6 and 7, respectively (Rand, 1986). The "unallocated" amounts represent the difference between the amount of estimated use and the total production data. According to Rand (1986), part of the unallocated use consists of unreported food refrigeration use and losses during storage, packaging, and transport.

Table 5. Estimated use patterns of chlorofluorocarbons in the USA[a]
(% total production)

Application	1975	1978	1981	1985
Aerosol propellants	50	24	<1	2
Refrigerants	28	39	46	39
Foam blowing agents	b	12	20	17
Solvents	5	11	16	14
Plastics and resins	10	b	7	14
Sterilant gas	b	b	2	2
Food freezant	b	b	1	1
Miscellaneous and export	7	14	7	11

[a] From: CMR (1975, 1978, 1981, 1986).
[b] Included in miscellaneous category.

Table 6. Estimated use patterns of CFC-11 in the USA and worldwide[a]
(excluding eastern European countries)

Use	USA	World
Blowing agent	71%	58%
Refrigeration	6%	3%
Aerosol	5%	31%
Miscellaneous	18%	8%

[a] From: Rand (1986).

In countries that signed the Montreal Protocol, the use of these chlorofluorocarbons will decline.

3.3.2 Release from use: controlled or uncontrolled

The release of CFC-11 and CFC-12 during use has caused the greatest concern environmentally because of their impact on ozone-depletion. During the mid-1970s, when aerosol propellant use was the major chlorofluorocarbon application, aerosols accounted for 75% of the immediate release of CFC-11 and CFC-12, while refrigerants and blowing agents accounted for 14% and 12%, respectively (Smart, 1980). Table 8 shows the estimated release of

Table 7. Estimated use patterns of CFC-12 in the USA and worldwide[a] (excluding eastern European countries)

Use	USA	World
Blowing agent	11%	12%
Mobile air-conditioning	37%	20%
Retail food refrigeration	4%	3%
Chillers	1%	1%
Home refrigerators	2%	3%
Aerosol	4%	32%
Miscellaneous	10%	7%
Unallocated	31%	22%

[a] From: Rand (1986).

these two chlorofluorocarbons in 1965, 1970, 1975, 1980, and 1985. CMA (1986) estimated that the total cumulative worldwide (with the exception of eastern European countries) release of CFC-11 and CFC-12 as a result of their use since 1931 amounted to 13.6 million tonnes in 1985.

Table 8. Worldwide production and release of CFC-11 and CFC-12 during use (thousands of tonnes)[a]

Year	Production	Release from:				Total release
		Refrigeration (hermetically-sealed)	Refrigeration (non-hermetic)	Blowing agent (closed-cell foam only)	Other sources	
1965	312.9	0.8	44.9	5.7	232.1	283.5
1970	559.2	1.2	68.5	16.3	420.7	506.7
1975	695.1	1.8	103.8	35.4	574.0	715.0
1980	639.8	2.6	156.1	65.0	359.7	583.4
1985	703.1	3.9	188.2	99.4	357.8	649.3

[a] From: CMA (1986).

4. ENVIRONMENTAL TRANSPORT, DISTRIBUTION, AND TRANSFORMATION

4.1 Transport between media

Because of the high vapour pressure of chlorofluorocarbons, the major transport medium is the atmosphere. For example, Lovelock (1972) found that CFC-11 concentrations in rural southern England and Ireland could be partly attributed to sources on the continent of Europe.

CFC-11 and CFC-12 introduced into aquatic systems will most likely volatilize to the atmosphere. Once in the troposphere, they will eventually diffuse into the stratosphere or be carried back to the earth through precipitation (Callahan et al., 1979).

Data pertaining to the adsorption of CFC-11 and CFC-12 onto soils and sediments are inconclusive (Callahan et al., 1979). However, the octanol/water partition coefficients of CFC-11 (log P = 2.53) and CFC-12 (log P = 2.16) (Hansch et al., 1975) indicate that adsorption onto organic particulates may be possible. In cases of significant sorption to soils, the volatilization of these compounds will be slower than in aquatic systems, though volatilization may still be the major transport process from soils.

4.2 Environmental transformation and degradation processes

4.2.1 Oxidation

No information is available concerning the oxidation of CFC-11 or CFC-12 in the aquatic environment under ambient conditions. These two chlorofluorocarbons are known to be relatively stable with respect to attack by hydroxyl radicals present in the troposphere (Lillian et al., 1975; US EPA, 1975; Cox et al., 1976; Hanst, 1978).

4.2.2 Hydrolysis

As a group, chlorofluorocarbons exhibit a low rate of hydrolysis compared with other halogenated compounds, and the rates of hydrolysis are greatly affected by temperature, pressure, and the presence of catalytic materials such as metals. Should hydrolysis of CFC-12 and possibly other chlorofluorocarbons occur, it would proceed at a negligible rate compared with the rate of volatilization and subsequent photodissociation.

4.2.3 Photolysis

4.2.3.1 Photochemistry

Atmospheric ozone prevents virtually all sunlight of wavelengths less than 290 nm from reaching the earth's surface. Since the wavelength of sunlight at altitudes below 50 km is greater than 280 nm, which is above the wavelength absorbed by chlorofluorocarbons (Doucet et al., 1973, 1974), there is no mechanism for direct photoalteration of these chemicals in the lower atmosphere.

4.2.3.2 Environmental transformation

CFC-11 and CFC-12 do not photodissociate in the troposphere, since they do not absorb radiation at wavelengths greater than 200 nm (Hanst, 1978). They eventually diffuse into the stratosphere (NRC, 1976; Hanst, 1978) where they are broken down by higher energy, shorter wavelength ultraviolet radiation (Jayanty et al., 1975; Rebbert & Ausloos, 1975; US EPA, 1975; Hanst, 1978; Isaksen & Stordal, 1981).

The photodissociation of CFC-11 and CFC-12 each result in the release of two chlorine atoms, since less energy is required to cleave the C-Cl bond than the C-F bond (Rebbert & Ausloos, 1975). According to Jayanty et al. (1975), the photolysis of CFC-11 in the presence of O_2 at 213.9 nm and 25 °C leads to the production of CFClO and, potentially, chlorine molecules (Cl_2), while the photolysis of CFC-12 under the same conditions leads to the production of CF_2O and Cl_2. Chlorine atoms, released by reactions such as these, are catalysts in the destruction of the stratospheric ozone layer (US EPA, 1975; Hanst, 1978; Ember, 1986; Zurer, 1988).

Isaksen & Stordal (1981) rationalized the ozone depletion by way of a cycle involving the intermediate formation of chlorine oxide (ClO). The net reaction for each turn of the cycle is as follows:

$$Cl + O_3 \rightarrow ClO + O_2$$
$$ClO + O \rightarrow Cl + O_2$$
$$\text{Net}\quad O + O_3 \rightarrow 2O_2$$

Other sequences involving ultraviolet radiation and radical species have also been proposed (Ember, 1986).

4.2.4 Biodegradation

No information on the biodegradability of the commercial chlorofluorocarbons is available (Su & Goldberg, 1976; Callahan et al., 1979).

4.3 Interaction with other physical, chemical, or biological factors

As indicated above, the commercial chlorofluorocarbons are relatively persistent in the environment because of their chemical stability, although their degree of persistence has not been determined with accuracy. The current best estimates for the average residence times in the atmosphere are 65, 110, 400, 90, 180, and 380 years for CFC-11, CFC-12, CFC-13, CFC-113, CFC-114, and CFC-115, respectively (NASA, 1986).

Assuming a troposphere-to-stratosphere turnover time (the time taken for 63% of troposphere air to diffuse into the stratosphere) of 30 years, tropospheric life-times of 65 and 110 years, respectively, would result in about 86% of tropospheric CFC-11 and CFC-12 eventually reaching the stratosphere. The effect of the transport of CFC-11 and CFC-12 from troposphere to stratosphere has been reviewed by NASA (1986). The addition of CFC-11 and CFC-12 to the atmosphere affects the climate in two ways. Firstly, these compounds have strong absorption bands in the atmospheric "window" region, that is from 7-13 μm. Therefore, both CFC-11 and CFC-12 will induce a "greenhouse" warming effect by direct absorption of terrestrial radiation. The second effect is due to the depletion of the stratospheric ozone layer. Mathematical modelling has shown that chlorofluorocarbons will reduce the ozone column. For instance, it has been calculated that a chlorofluorocarbon growth rate of 3% per year would lead to a 10% ozone depletion within 70 years (NASA, 1986). Changes of that magnitude, or even smaller ones, could have important biological consequences (sections 6 and 9.4). Additions of chlorofluorocarbons to the atmosphere are also predicted to modify the vertical distribution within the ozone column. As a result of the unique regional meteorology and the presence of chlorine radicals in the Antarctic stratosphere, stratospheric ozone reductions of 30-50% have been recently observed there during the austral spring.

The reduction of stratospheric ozone affects the surface in two ways:

- directly by increasing the penetration of ultraviolet B radiation (290-320 nm);
- indirectly by enhancing the global warming effects and altering climatic conditions.

Bioconcentration and bioaccumulation

Dickson & Riley (1976) have found CFC-11 at levels of 0.6-28 µg/kg (dry weight basis) in various organs of fish and molluscs. These levels, however, do not necessarily indicate a potential for bioaccumulation.

Neely et al. (1974) suggested that bioaccumulation is directly related to the octanol/water partition coefficient (P) of the compound. The experimentally determined log octanol/water partition coefficients (log P) of CFC-11 and CFC-12 (see section 4.1.1) indicate that the bioaccumulation potential in organisms is low.

5. ENVIRONMENTAL LEVELS AND HUMAN EXPOSURE

5.1 Environmental levels

5.1.1 Air

Singh et al. (1979) collected *in situ* air samples aboard a US Coast Guard vessel that sailed the Pacific Ocean from Oakland, California, USA (37 °N) to Wellington, New Zealand (42 °S). Tyson et al. (1978) made measurements at latitudes from 74 °N to 62 °S as part of a 1976 NASA Latitude Survey Mission between Alaska, USA, and New Zealand. The results of their monitoring are summarized in Table 9.

Table 9. Global distribution of chlorofluorocarbons in the troposphere (ng/m^3)[a]

Chlorofluorocarbon	Northern hemisphere		Southern hemisphere		Reference
	Mean	Standard deviation	Mean	Standard deviation	
CFC-11	747.5 (113)	75.3 (13.4)	668.8 (119)	65.8 (11.7)	Singh et al. (1979)
	741.8 (132)	50.6 (9)	696.9 (124)	33.7 (6)	Tyson et al. (1978)
CFC-12	1138.5 (230)	126.2 (25.5)	1039.5 (210)	124.2 (25.1)	Singh et al. (1979)
	1079.1 (218)	54.4 (11)	821.7 (166)	39.6 (8)	Tyson et al. (1978)
CFC-113	145.9 (19)	26.8 (3.5)	138.1 (18)	23.8 (3.1)	Singh et al. (1979)
CFC-114	83.9 (12)	13.3 (1.9)	69.9 (10)	9.1 (1.3)	Singh et al. (1979)

[a] Figures in brackets are in parts per trillion (by volume).

The increased use of chlorofluorocarbons on a worldwide basis has resulted in an increase in the global levels of these compounds. The two most abundant chlorofluorocarbons in the atmosphere are CFC-11 and CFC-12 (Guicherit & Schulting, 1985). The annual growth rates in the 1980s appear to be slower than the growth rates in the 1970s (Rasmussen et al., 1981). The annual rate of increase in CFC-11 global levels during the period 1975-1980 was 8-12% (Rasmussen et al., 1981; Fraser et al., 1983;

Singh et al., 1983), whereas it was 6-7% during 1980-1981 (Brice et al., 1982; Cunnold et al., 1983b; Prinn et al., 1983; Rasmussen & Khalil, 1986). Similarly, although the average annual growth rate for global levels of CFC-12 during 1975-1980 was 8-9% (Rasmussen et al., 1981; Singh et al., 1983), it was only 6% in 1980 (Cunnold et al., 1983b; Prinn et al., 1983; Rasmussen & Khalil, 1986). Both CFC-11 and CFC-12 showed an accumulative increase of about 60% during the decade 1975-1985 (Rasmussen & Khalil, 1986).

Data on the atmospheric concentrations of chlorofluorocarbons are shown in Table 10. Measurements of atmospheric chlorofluorocarbon concentrations up to an altitude of 6 km did not reveal any significant concentration changes with increasing altitude (Rasmussen & Khalil, 1982, 1983, 1986; Robinson et al., 1983). Hunter-Smith et al. (1983), Rasmussen & Khalil (1983), and Singh et al. (1983) studied the latitudinal variation in chlorofluorocarbon concentrations between the northern and southern hemisphere and reported inter-hemispheric contrasts (ratio of concentration between northern and southern hemispheres) of 1.08 for CFC-11, 1.07-1.08 for CFC-12, 1.10-1.25 for CFC-113, and 1.08 for CFC-114. Table 10 reveals that the concentrations of chlorofluorocarbons are higher in urban areas than in remote areas, this being the result of local emission sources. The urban concentrations of chlorofluorocarbons (CFC-11 and CFC-12) in the People's Republic of China, with the exception of Beijing, are the same as background levels in the USA. This is probably due to the less extensive use of these compounds in urban areas in China (Rasmussen et al., 1982).

Median concentrations of the most abundant compounds, CFC-11 and CFC-12, in several urban/suburban areas and rural/remote areas in the USA are reported in Table 10 (Brodzinsky & Singh, 1982). These measurements were made from 1972 to 1980, the median year being 1975. Median concentration values of 1090 and 3420 ng/m^3 for CFC-11, and 1630 and 5700 ng/m^3 for CFC-12, in rural/remote and urban/suburban areas, respectively, were projected for 1985, assuming that the average annual growth rate for both compounds would be 5% (NASA, 1986). Assuming that an individual inhales 20 m^3 air/day, the total inhalation exposure (CFC-11 plus CFC-12) in 1985 would be 54 or

Table 10. Some worldwide measurements of the atmospheric concentrations of chlorofluorocarbons

Location	Year	Concentration of chlorofluorocarbons (ng/m^3)				Reference
		CFC-11	CFC-12	CFC-113	CFC-114	
Barbados						
Ragged Point	1980	NR	1499	NR	NR	Cunnold et al. (1983b)
	1985	1313	2012	NR	NR	NASA (1986)
Samoa (American)						
Point Matatula	1980	NR	1433	NR	NR	Cunnold et al. (1983b)
United Kingdom						
Harwell	1980	1342	NR	NR	NR	Brice et al. (1982)
Adrigole, Ireland	1980	NR	1564	NR	NR	Cunnold et al. (1983b)
USA						
Phoenix, Arizona	1979	1423	NR	1192	NR	Singh et al. (1981)
Los Angeles, California	1979	2700	NR	2376	NR	Singh et al. (1981)
Oakland, California	1979	1365	NR	381	NR	Singh et al. (1981)
USA rural/remote (median concentration)	1973-1980	685	1911	241	64	Brodzinsky & Singh (1982)
USA urban/suburban (median concentration)	1972-1980	1199	3521	1324	199	Brodzinsky & Singh (1982)
Pacific Northwest	1980	1073	1620	132	NR	Rasmussen et al. (1981)
Northern hemisphere	1978	919	1378	101	NR	Rasmussen & Khalil (1982)
Northern hemisphere	1978	1062	1534	179	100	Singh et al. (1983)
Southern hemisphere	1978	845	1283	93	NR	Rasmussen & Khalil (1982)
Southern hemisphere	1978	982	1418	164	92	Singh et al. (1983)
Arctic	1982	1174	1780	175	NR	Rasmussen & Khalil (1983)
Arctic haze	1979	1097	1633	NR	NR	Khalil & Rasmussen (1983)
South Pole	1980	948	1428	86	NR	Rasmussen et al. (1981)
Over Atlantic Ocean	1981	1056	NR	NR	NR	Brice et al. (1982)
Global average	1980	959	NR	NR	NR	Fraser et al. (1983)

NR = not reported.

182 µg/day in rural/remote or urban/suburban areas of the USA, respectively. Using the 1985 data from Ragged Point, Barbados, as a basis, the inhalation for combined CFC-11 and CFC-12 in rural/remote areas in late 1985 would have been 66 µg/day.

5.1.2 Water

Singh et al. (1979) measured CFC-11 and CFC-12 concentrations in 1977 at various locations in the Pacific Ocean. The average surface concentration of CFC-11 was 0.13 ($\pm$ 0.006) ng/litre, while the CFC-12 concentration was 0.28 ($\pm$ 0.15) ng/litre. The average concentrations at a depth of 300 m were 0.06 and 0.21 ng/litre for CFC-11 and CFC-12, respectively. The concentrations of CFC-11 and CFC-12 at various locations in the eastern Pacific Ocean (surface waters) during 1979-1981 were 0.22 and 0.25 ng/litre (Singh et al., 1983), in Greenland Sea surface water in 1982 were 0.61 and 0.21 ng/litre (Bullister & Weiss, 1983), and in Japanese coastal waters were 0.20-0.54 and 0.19-0.33 ng/litre, respectively (Tomita et al., 1983).

Samples of water from Lake Ontario analysed for volatile halocarbon contaminants contained mean concentrations for CFC-11 and CFC-12 of 249 and 572 ng/litre, respectively (Kaiser et al., 1983). An alluvial aquifer in Southington, Connecticut, USA, adjacent to a solvent-recovery operation was analysed in 1980 for volatile organic compounds, but CFC-12 was not detected (detection limit not specified) in water obtained from various depths (Hall, 1984). CFC-11 and CFC-12 have been detected in surface snow and rainwater in Alaska (Su & Goldberg, 1976). The detection of chlorofluorocarbons in drinking-water has not been reported.

5.1.3 Food and other edible products

With the exception of a few scattered reports (section 4.4), chlorofluorocarbons have not been measured in food.

5.2 Occupational exposure

Information on occupational exposure is summarized in section 9.2.

6. ECOLOGICAL EFFECTS OF STRATOSPHERIC OZONE DEPLETION

6.1 Introduction

Speculation on the possibility of stratospheric ozone reduction first appeared in the early 1970's and focused on the consequences of large quantities of nitrogen oxides being injected into the upper atmosphere by supersonic aircraft flying at high altitudes. Other sources of nitrogen oxides originating from the earth's surface were also considered. These concerns gradually diminished, because the quantities of nitrogen oxides likely to be involved were insufficient to cause a serious threat to the ozone layer. However, concern over halogen pollution of the upper atmosphere arose during the mid-1970s (section 4.2.3). The halogens of immediate concern were chlorine and bromine. The main source for chlorine is chlorofluorocarbons, which are released worldwide from such sources as aerosol spray cans, certain plastic foams, refrigerators, and refrigerative air conditioners.

Many gases emitted as a result of industrial and agricultural activities can accumulate in the Earth's atmosphere and ultimately contribute to alterations in the vertical distribution and concentrations of stratospheric ozone. Among the most important are those trace gases that have long residence times in the atmosphere. This allows accumulation in the troposphere and a gradual upward migration of the gases into the stratosphere where they contribute to depletion of stratospheric ozone. The atmospheric and chemical processes involved are extremely complex (US EPA, 1987a). Trace gases of particular concern include certain long-lived chlorofluorocarbons, such as CFC-11, CFC-12, and CFC-113 (for atmospheric residence times see section 4.3). Since the transport of these gases to the stratosphere is slow, their residence times there are long, and the removal processes are slow, any effect on stratospheric ozone already seen is probably the result of anthropogenic emissions of these gases several decades ago. Those gases already in the atmosphere will continue to exert stratospheric ozone depletion effects well into the next century.

The atmospheric models that predict future ozone depletion are in a continual process of refinement. Over the years, predicted decreases in stratospheric ozone have ranged from 4 to 18%, based on the stratospheric concentrations of chlorine expected from the 1974 levels of CFC-11 and CFC-12 emissions. However, it has gradually been realized that other gases will influence column ozone and that the size and direction of the predicted change in total ozone during the next century depend critically on the assumption of the multiple trace-gas scenarios. Many of the modelling scenarios tended to assume relatively uniform rates of ozone layer reduction widely distributed above all regions of the Earth. However, areas of distinctly greater depletion (ranging from 15 to 40% in recent years) have been identified over the South Polar region during September to November of each year. The evidence suggests a likely gradual expansion of this "Antarctic Ozone Hole" ultimately to extend beyond the South Polar region, possibly coming to reach over more heavily populated areas of the Southern Hemisphere. Similarly, it is considered likely that an analogous, though less intense, zone of upper level ozone reduction will occur over the North Polar region and expand over populated areas of the Northern Hemisphere.

Although ozone constitutes a very small proportion of the stratosphere, it plays a major role in protecting life on this planet. The result of changes in the density of the total ozone column could, therefore, be far-reaching. The natural distribution of ozone in the Earth's atmosphere, concentrated most heavily in a diffuse layer in the stratosphere, is crucial in helping to protect human beings, other biological systems, and man-made materials from the harmful effects of certain wavelengths of sunlight. Stratospheric ozone exerts its beneficial effects by absorbing ultraviolet radiation in the 200- to 320-nm range, allowing only reduced amounts of UV-B radiation (280- to 320-nm waveband) to penetrate to the Earth's surface. In addition, the vertical distribution of stratospheric ozone and relative dryness of the air in the stratosphere help to maintain the radiative balance of the Earth. Depletion of the stratospheric ozone layer can, therefore, be expected to lead to damaging effects on human health and the environment (i) directly by increased penetration of UV-B radiation to the Earth's surface and (ii) indirectly through the influence of changes in the vertical distribution of stratospheric ozone and water vapour that contribute to global warming effects and altered climatic conditions. The possibility of increased exposure to solar UV-B radiation is a particular cause for concern because of its effect on humans, other animals, plants, certain manufactured materials, and photochemical smog production. Most of the known biological effects of UV-B radiation are damaging. Detailed discussions of evolving concern about stratospheric ozone depletion and assessment of the scientific base underlying such concern can be found in several recent national and international expert work group reports or symposia (e.g., US EPA, 1987a; Schneider et al., 1989; WMO/Canada DOE, 1989). The following sections summarize key points from such sources and discuss their implications for the development of effective international efforts to cope with ozone layer depletion.

6.2 Terrestrial plants

Increased UV-B irradiation of the Earth's surface due to ozone layer depletion can be expected to have a negative impact on both terrestrial and aquatic biota. In assessing the impact of increased exposure to UV-B radiation for crops and terrestrial ecosystems, it must be recognized that existing knowledge is in many ways deficient. The effects of enhanced levels of UV-B radiation have been studied in only a few representative species from some of the major terrestrial ecosystems. Most knowledge has been derived from studies that focused upon agricultural crops and were conducted at mid-latitudes. Despite uncertainties resulting from the complexities of field experiments, the available data suggest that crop yields are vulnerable to increased levels of solar UV-B radiation. Unlike drought or other geographically isolated stresses, stratospheric ozone depletion would affect all areas of the world, including ecosystems whose UV-B sensitivity has not been investigated.

Out of more than 200 species and cultivars screened for UV tolerance, about two-thirds have been found to be sensitive. Most tests were done in controlled environments with UV radiation from artificial sources. The UV sensitivity was usually exaggerated when compared to results obtained by exposure to solar radiation in the field. The most sensitive plant groups include crops related to peas and beans, melons, mustard, and cabbage, but there are large differences in sensitivity between the various crops studied in the field (US EPA, 1987b). In general, UV radiation causes reduced leaf and stem growth, lower total dry weight, and lower photosynthetic activity in sensitive cultivars (Tevini & Iwanzik, 1986). These results were corroborated in an experiment simulating a 25% enhancement of solar UV-B radiation (equivalent to 12% ozone reduction), where UV-B exposure was controlled by an artificial ozone filter at a high altitude and at a southern latitude (Tevini et al., 1986). Members of the grass family were generally less sensitive (with some notable exceptions), possibly due to protective abilities such as photorepair or production of screening pigments (Beggs et al., 1986).

The large variation in sensitivity that exists among cultivars within each crop species suggests that some degree of UV tolerance must be present in the existing gene pool. The genetic basis for differences in UV-B sensitivity is not fully understood. However, it is possible that selective crop breeding might help mitigate some of the potentially deleterious effects (Teramura, 1983).

In addition to other factors, the quality of crop yield may be reduced by increased levels of UV-B radiation. Changes in crop quality have not been specifically examined in many studies, but reduced quality has been noted in certain cultivars of tomato, potato, sugar beet, and soybean. The protein and oil content of specific cultivars of soybean seeds were reduced by up to 10% when plants were exposed to UV levels equivalent to a 25% ozone depletion (US EPA, 1987b).

Increased levels of UV-B radiation may also affect forest productivity. Only limited data are available on coniferous species, but in studies by Sullivan & Teramura (1988) about one-half of the species of seedlings were adversely affected by UV-B radiation. In loblolly pine seedlings, growth and photosynthesis were reduced in field studies simulating a 40% ozone reduction (Teramura &

Sullivan, 1988). However, extrapolation from the results of seedling studies to forested ecosystems is not possible, nor is interpolation of predicted results at exposure levels simulating a lower level of ozone reduction.

The existing data also suggest that increased UV-B radiation will modify the distribution and abundance of plants, and potentially change ecosystem structure as a result of an alteration of the competitive balance between different species. Even small changes in competitive balance over a period of time can result in large changes in community structure and composition (Gold & Caldwell, 1983). The shift in competitive balance may occur in response to subtle changes in plant growth, without large changes in fundamental physiological processes such as photosynthesis (Beyschlag et al., 1988). The alteration of the competitive balance of species is a dynamic process affected by the competing species and their immediate environment. Unfortunately, neither a quantitative nor a qualitative prediction of how these ecosystems might be altered can be determined from the current knowledge base.

6.3 Aquatic organisms

Various experiments have demonstrated that UV-B radiation causes damage to fish larvae and juveniles, shrimp larvae, crab larvae, copepods, and plants essential to the marine food web. These damaging effects include decreased fecundity, growth, survival, and other reduced functions in these organisms (Worrest, 1982; US EPA, 1987c). Evidence indicates that ambient solar UV-B radiation, although not nearly as important as light, temperature, or nutrient levels, is currently an important limiting ecological factor, and that even small increases in UV-B exposure could result in significant ecosystem changes (Damkaer, 1982).

Effects induced by solar UV-B radiation have been measured to a depth of more than 20 metres in clear waters and more than five metres in less clear water. The euphotic zone (i.e. water depth with levels of light sufficient for positive net photosynthesis) is frequently taken as the water column that reaches down to the depth at which photosynthetically active radiation is reduced by

99%. In marine ecosystems, UV-B radiation penetrates approximately the upper 10% of the marine euphotic zone before it is reduced by 99% of its surface irradiance. Penetration of UV-B radiation into natural waters is a key variable in assessing the potential impact of this radiation on any aquatic ecosystem (US EPA, 1987c).

In marine plant communities a change in species composition rather than a decrease in net production would be the probable result of increased UV-B exposure (Worrest, 1983). A change in community composition at the base of food webs may produce instabilities within ecosystems that could affect higher trophic levels (Kelly, 1986). The generation time of marine phytoplankton is in the range of hours to days, whereas the potential increase in ambient levels of solar UV-B irradiance will occur over decades. The question remains as to whether the gene pool within species is capable of adapting during this relatively gradual (relative to the generation time of the target organisms) change in exposure to UV-B radiation. There is evidence that a decrease in column ozone abundance could diminish the near-surface season of invertebrate zooplankton populations. For some zooplankton, the time spent at or near the surface is critical for food gathering and breeding. Whether these populations could endure a significant shortening of the surface season is unknown (Damkaer et al., 1980).

The direct effect of UV-B radiation on edible fish larvae closely parallels the effect on invertebrate zooplankton. More information is required on seasonal abundances and vertical distributions of fish larvae, vertical mixing, and penetration of UV-B radiation into appropriate water columns before effects of exposure to solar UV-B radiation can be predicted. However, in one study involving anchovy larvae, it was calculated that a 20% increase in UV-B radiation (which would accompany a 9% depletion of total column ozone) would result in the death of about 8% of the annual larval population (Hunter et al., 1982). This one study was performed in the laboratory, and even the control animals had significant mortality at the end of the normal larval period. This highlights the need for caution when trying to extrapolate conclusions to natural conditions when those conclusions are based on results from laboratory studies.

In many countries marine species supply more than 50% of the dietary protein, and in developing countries this percentage is often higher. Research is needed to improve our understanding of how stratospheric ozone depletion could influence the world food supply. However, effective steps to minimize stratospheric zone depletion cannot await the outcome of such research.

6.4 Research needs

Future work concerning UV-B effects on terrestrial ecosystems must proceed on a broad front. Sensitivity screenings and dose-response studies must expand to include representative species from a wider range of ecosystem types and a wider range of plant types within ecosystems of particular interest. Knowledge of species sensitivities and their geographic ranges can then be combined with information on current and projected levels of UV-B in order to identify areas of greatest concern. An understanding of how sensitivity to UV-B is affected by other environmental factors will aid in this process. Additional work at the biochemical level is needed to clarify interactions of UV-B radiation and plant metabolism as well as the nature of effects of UV-B radiation on pests and pathogens.

Ultimately, the information gathered in field and laboratory studies must be put into the context of ecosystem properties, including primary productivity, nutrient cycling, resistance to disturbance, and the capacity to recover from disturbance. Efforts are clearly needed to integrate what is known about the influences of elevated UV-B irradiance on plants with what is known about plant stress associated with other human-induced changes in the environment.

In order to quantify the effects on marine systems of UV-B radiation on an ocean-wide basis, there is a need for additional data on the penetration of UV-B radiation as a function of water mass, concentration of particulates, and presence of plankton. These data must be combined with accurate measurements of total incident radiation, as a function of angle of incidence and time, to arrive at reliable estimates of both total UV-B radiation dose and dose rate.

There is a clear need to measure fish-larval sensitivity to UV-B radiation for many resource species, refine the links between exposure of primary producers to UV-B radiation and effects on fish, assess the impact of food-web changes on fish yield, and delineate the mitigating mechanisms available to the organism.

Studies on changes in population size and diversity as a result of stress would provide insights for predictions of the effects of UV-B increases in a given ecological niche (Worrest et al., 1978, 1981a,b; Worrest, 1983). Data describing changes resulting from environmental stress, such as contamination from toxic substances or temperature change, could be combined with data on the efficiency of energy conversion between trophic levels, upon which a resource species relies, to estimate the potential reduction in fish catch. To narrow the reliability limits of such predictions, field investigations into the resiliency of affected populations are required.

There is a paucity of information on the impact of UV-B radiation on marine resource species. The fact that dose-response sensitivity data exist for only a few species greatly impedes our ability to extrapolate to an overall assessment of the risk to marine fisheries. It is important to be able to translate known intracellular cause-and-effect relationships of UV damage to effects on simple or single-celled organisms and to population effects.

Knowledge of adaptive or protective mechanisms by which marine organisms minimize the effects of increased UV-B radiation in the ocean's surface layers is lacking. No avoidance mechanisms specific to UV-B radiation have been described for marine organisms, although avoidance mechanisms to visible light may lessen the impact of the concurrent UV radiation. While pigmentation occurs extensively in marine organisms, the degree to which it contributes to UV-B protection is unknown.

The time scale of adaptation or repair, compared to the time scale of increased UV-B radiation, is an important factor. Are genetic mechanisms sufficient to obviate the negative impacts? Do they affect competitor species over similar time scales? What organisms are pre-disposed to environmental (i.e. non-genetic) protective behaviour?

These questions must be addressed as part of the framework of risk assessment.

7. KINETICS AND METABOLISM

7.1 Absorption

Chlorofluorocarbon propellants and solvents may present a hazard for human beings by inhalation, ingestion, and dermal absorption. However, because of the physical properties and uses of these compounds, inhalation is the most common route of entry, and exhalation is the most significant route of elimination.

Information concerning chlorofluorocarbon absorption has been obtained in two types of studies:

- chlorofluorocarbon retention in the lungs;
- chlorofluorocarbon blood levels after inhalation.

The relative amounts of CFC-11, CFC-12, CFC-113, and CFC-114 absorbed by human beings have been measured in breath-holding studies (Paulet & Chevrier, 1969; Morgan et al., 1972). Retention was measured using radioisotopically marked chlorofluorocarbons by subtracting the radioactivity exhaled 30 min after inhalation from the amount of radioactivity inhaled with a single breath. In terms of absorption the following order was obtained: CFC-11 $\simeq$ CFC-113 > CFC-114 $\simeq$ CFC-12, with retentions of 23%, 19.8%, 12.2%, and 10.3%, respectively. Shargel & Koss (1972) exposed dogs to an equal weight mixture of CFC-11, CFC-12, CFC-113, and CFC-114, and obtained similar results.

In other studies, human volunteers (Aviado & Micozzi, 1981) and dogs (Azar et al., 1973) were exposed to CFC-11 at a concentration of 5710 mg/m^3 (1000 ppm) and for a period of 8 h or 10 min, respectively. The blood levels in the human volunteers were 4.69 µg/ml and in the dogs 6.5-10 µg/ml. According to a mathematical model developed for the description of the pharmacokinetics, 77% of the dose applied was absorbed.

Azar et al. (1973) determined the corresponding data for CFC-12 in beagle dogs. After an exposure to 5030 mg/m^3 (1000 ppm) for a period of 10 min, 1.1 µg/ml was found in the arterial blood and 0.4 µg/ml in the venous system. At higher concentrations, the arterial and venous concentrations were similar. Trochimowicz et al. (1974) found, under similar conditions (1000 ppm, 1 min inhalation period), that the blood level in dogs for CFC-113 was 2.7 µg/ml (arterial) and 1.9 µg/ml (venous), and for CFC-114 0.4 µg/ml and 0.2 µg/ml, respectively.

In a study by Angerer et al. (1985), three volunteers were exposed to a CFC-11 concentration of 3750 mg/m^3 (657 ppm). The average value of pulmonary retention was 18.9%. CFC-11 levels in alveolar air and blood were 3066 mg/m^3 (537 ppm) and 2.8 µg/ml, respectively.

Further absorption and elimination data from CFC-11 and CFC-12 atomizer administrations indicated that, while CFC-11 is more readily absorbed by mammals (including humans) than CFC-12, the degree of preferential absorption may vary among individuals (Dollery et al., 1970; Allen & Hanburys Ltd, 1971; Paterson et al., 1971; Shargel & Koss, 1972). Similar information on the different absorption rates has been obtained from other studies. Chlorofluorocarbons were administered to dogs for 5 min at fixed concentrations between 0.3 and 10 vol % in the inspired air. The blood concentrations determined up to 60 min after exposure indicated that CFC-11 is more readily absorbed than CFC-12 or CFC-114 (Clark & Tinston, 1972a).

The results of Adir et al. (1975) and Brugnone et al. (1984) provide additional evidence that CFC-11 is absorbed to a greater extent than CFC-12 in dogs and rabbits. The absorption data correlate well with the liquid/gas partition coefficients for these compounds in whole blood, serum, and olive oil shown in Table 11.

CFC-12 was absorbed 4 times more readily than CFC-114 in a study by Rauws et al. (1973) in which rats were exposed to a mixture of CFC-11, CFC-12, and CFC-114 (weight ratio of 1:2:1). A similar pattern was also seen in monkeys by Taylor et al. (1971). In each instance, the ratio of CFC-12 to CFC-114 in arterial blood was higher than the ratio of exposure concentrations, indicating that CFC-12 was slightly more readily absorbed than CFC-114.

The available data on chlorofluorocarbon uptake indicate that chlorofluorocarbons can be absorbed across the

Table 11. Partition coefficients of various chlorofluorocarbons

Compound	Whole blood[a] (rat)	Whole blood[b] (human)	Serum[c] (human)	Olive oil[c]
CFC-11	1.4	0.87	0.9	27
CFC-12	0.2	0.15	0.2	3
CFC-113			0.8	32
CFC-114		0.15	0.2	5

[a] From: Allen & Hanburys, Ltd (1971).
[b] From: Chiou & Niazi (1973).
[c] From: Morgan et al. (1972).

alveolar membrane, gastro-intestinal tract, the skin, and internal organs. Following inhalation, they are absorbed rapidly by the blood. Blood-tissue absorption is probably the rate-limiting step. After an initial, rapid blood level stabilization, chlorofluorocarbons are still absorbed by body tissues and continue to enter the body.

7.2 Distribution

Allen & Hanburys, Ltd. (1971) found in mice that both CFC-11 and CFC-12 are taken up by heart, fat, and adrenal tissue after 5-min inhalation exposures. CFC-11 is concentrated from the blood to the greatest extent in the adrenals followed by the fat, then the heart. A similar, though less pronounced, pattern is evident for CFC-12 but CFC-11 is absorbed and concentrated in all of these tissues to a much greater extent than CFC-12. Paulet et al. (1975) noted that both CFC-11 and CFC-12 are distributed to the cerebrospinal fluid of dogs after inhalation exposure.

Following inhalation exposures lasting 7-14 days, Carter (1970) noted distribution patterns for CFC-113 in rats that were qualitatively similar to those noted for CFC-11 and CFC-12 by Allen & Hanburys, Ltd. (1971). The major difference from the CFC-11 and CFC-12 results was that almost all of the CFC-113 concentration occurred in the fat, while adrenal levels were relatively low and even decreased as exposure continued (it should be emphasized that the exposures to CFC-11 and CFC-12 were for only

5 min). The other organ levels did not change significantly from a 7-day to a 14-day exposure, which is consistent with the idea that such concentrations will stabilize as equilibria between ambient air concentration, blood level, and tissue levels are reached. In rats and guinea-pigs, shortly after exposure to CFC-113, Furuya (1979) noted the following tissue distribution in decreasing order: fat, brain, liver, kidney, heart, lung, muscle, and blood.

In summary, chlorofluorocarbons are rapidly absorbed after inhalation and are distributed by blood into practically all tissues. Relatively high concentrations are found in fat, but also in organs with good blood supply.

7.3 Metabolic transformation

Of the nine chlorofluorocarbons reviewed in this document, some data regarding metabolism exists only for CFC-11, CFC-12, CFC-112a, and CFC-113a.

Cox et al. (1972a) found no evidence of reductive dehalogenation of CFC-11 in microsomal preparations from rats, chickens, or other species. However, the reductive dechlorination *in vitro* of CFC-11 to HCFC-21 by rat liver microsomes was reported by Wolf et al. (1978). *In vitro* metabolism studies suggested that CFC-112a and CFC-113a can be metabolized by reductive dechlorination and that the reaction is catalyzed by cytochrome P-450 from rat liver microsomes. However, no metabolites of either compound were identified (Salmon et al., 1981, 1985; Nastainczyk et al., 1982a,b).

Published studies on *in vivo* metabolism exist only for CFC-11 and CFC-12. Eddy & Griffith (1971) administered ^{14}C-labelled CFC-12 to rats by the oral route and reported a small amount of metabolism. About 2% of the total dose was exhaled as $^{14}CO_2$ and 0.5% was excreted in urine. CFC-12 and/or its metabolites were no longer detectable in the body 30 h after administration.

Blake & Mergner (1974) exposed beagle dogs for 6-20 min to CFC-11 (5710 to 28 550 mg/m^3; 1000 to 5000 ppm, v/v) or CFC-12 (40 240 to 60 380 mg/m^3; 8000 to 12 000 ppm, v/v) containing up to 180 µCi of ^{14}C-chlorofluorocarbon. Virtually all the administered chlorofluorocarbon was recovered in exhaled air within one hour with either material. Only traces of radioactivity were found in urine or exhaled CO_2 and may have represented unavoidable radiolabelled impurities rather than metabolites. The authors concluded that less than 1% of either CFC-11 or CFC-12 is metabolized after inhalation. The preceding results were essentially confirmed in human volunteers by the same authors (Mergner et al., 1975). Radiolabelled CFC-11 (571 mg/m^3; 100 ppm) and CFC-12 (503 mg/m^3; 100 ppm) were given by inhalation to one male and one female volunteer for 7-17 min. As was the case in dogs, little or no biotransformation of either chlorofluorocarbon was observed. Total metabolites were equal to, or less than, 0.2% of the administered dose.

The results of the preceding studies suggest that CFC-11 and CFC-12 are metabolized to a very small extent, if at all, in mammals following brief inhalation exposures.

7.4 Elimination and excretion in expired air, faeces, and urine

Regardless of the route of entry, chlorofluorocarbons appear to be eliminated almost exclusively through the repiratory tract. Little, if any, chlorofluorocarbon or metabolite has ever been reported in urine or faeces (Matsumoto et al., 1963; Blake & Mergner, 1974; Mergner et al., 1975).

7.5 Retention and turnover

When exposure is terminated, the more readily absorbed compounds are retained longer. The retention of chlorofluorocarbons after inhalation follows the same order as the amount absorbed during exposure:

$$CFC-11 \simeq CFC-113 > CFC-114 \simeq CFC-12$$

In human studies designed to mimic exposures to chlorofluorocarbons from atomizers, the initial blood half-lives for CFC-11 were in the range of 6 seconds to 1 min (Paterson et al., 1971).

In one study, volunteers exposed to CFC-11 at 3751 mg/m^3 (657 ppm) for 150-210 min showed half-lives for the initial and second phases of elimination from venous blood of 11 min and 1 h, respectively (Angerer et al., 1985). Half-lives for the initial and second phases of CFC-11 elimination in alveolar air were 7 min and 1.8 h, respectively (Angerer et al., 1985). Average pulmonary retention at an apparent steady state after 1 h of exposure was 18.2%. Similarly, the data of Brugnone et al. (1984) indicate a pulmonary retention of 19% for CFC-11 and 18% for CFC-12 in workers during occupational exposure.

Studies in which dogs were administered CFC-11 or CFC-12 by intravenous infusion indicated that the elimination of CFC-11 and CFC-12 from venous blood was triphasic (Niazi & Chiou, 1975, 1977). A 3-compartment model was proposed with initial, intermediate, and terminal half-lives of 3.2, 16, and 93 min for CFC-11 and 1.47, 7.95, and 58.50 min for CFC-12. Adir et al. (1975) also fitted their venous blood elimination data to a 3-compartment model. Estimates of half-lives for the terminal phases of CFC-11 elimination were 6.30 and 24.75 min for two human volunteers and 13.86-21 min (mean, 18.34) for four dogs. For the terminal phases of CFC-12 elimination, the half-lives were 9.63 min for one human volunteer and 8.45-11.35 min (mean, 9.90) for three dogs.

In dogs exposed to CFC-11 by atomizers, the initial and terminal half-lives in venous blood were at 0.6 and 4.03 min, respectively (McClure, 1972). The terminal half-life of 80 min in dogs after exposures to ambient CFC-11 concentrations of 2, 5, and 7.5% (Amin et al., 1979) is close to the terminal half-lives reported by Niazi & Chiou (1975).

Reinhardt et al. (1971a) conducted retention studies on CFC-113 in human volunteers over occupationally relevant periods. They measured the chlorofluorocarbon concentration in the expired air of volunteers exposed to 3835 mg/m^3 (0.05%) or 7670 mg/m^3 (0.1%) for 3 h in the morning and 3 h in the afternoon. Although there was no indication of chlorofluorocarbon accumulation, detectable levels were retained overnight in four cases at 3835 mg/m^3 and in 14 cases at 7670 mg/m^3. In one instance, there was a detectable level on a Monday morning following a final exposure to 7670 mg/m^3 (0.1%) on the previous Friday.

7.6 Reaction with body components

Lessard & Paulet (1985) concluded that simple dissolution of CFC-12 in the lipid layer of biological membranes with ensuing alteration of membrane configuration may account for its anaesthetic effect and some of its cardiac effects. Young & Parker (1972), however, suggested that CFC-12 is bound to the hydrophilic areas of various phospholipids and that potassium chloride may stop adrenaline-induced arrhythmia in hearts sensitized by CFC-12 by displacing the CFC-12 molecule held by the phospholipid.

CFC-11 has been shown to bind *in vitro* to liver microsomal protein and lipid (Uehleke et al., 1977; Cox et al., 1972a,b) and to cytochrome P-450 (Cox et al., 1972a,b; Wolf et al., 1977, 1978). Vainio et al., (1980) also demonstrated binding of CFC-113 to cytochrome P-450. In view of the very low liver toxicity potential of CFC-11 and CFC-113, the toxicological significance of the P-450 binding is unknown.

8. EFFECTS ON EXPERIMENTAL ANIMALS AND *IN VITRO* TEST SYSTEMS

8.1 Single exposures

8.1.1 Acute inhalation toxicity

A number of chlorofluoromethanes and chlorofluoroethanes have been tested for acute inhalation toxicity in laboratory animals. Because most of the information is of limited importance for a quantitative risk assessment, it will not be discussed in detail. The data are presented in Table 12.

Of the fully halogenated chlorofluoromethanes, CFC-12 and CFC-13 show extremely low acute inhalation toxicity. CFC-11 also has low acute inhalation toxicity, lethal concentrations being in the range of 571-1427 g/m^3 (100 000-250 000 ppm).

Within the chlorofluoroethanes, CFC-114 and 115 seem to be of an extremely low acute toxicity, followed by CFC-113 and CFC-112.

The symptomatology of acute intoxication is characterized by central nervous system effects and secondary effects on the cardiovascular and respiratory systems.

8.1.2 Acute oral toxicity

Very little information is available on the acute oral toxicity of chlorofluorocarbons. The lethality data for some chlorofluorocarbons are summarized in Table 13. With the exception of a slight increase in liver weight following exposure to CFC-112 and CFC-112a at 25 000 mg/kg, no gross or histological abnormalities were noted by Clayton (1966).

8.2 Short-term exposures

In this monograph, short-term exposures are defined as those involving repeated daily exposure up to 90 days and long-term studies as those longer than 90 days (see 8.4).

Table 12. Acute inhalation toxicity of fully halogenated chlorofluorocarbons

Compound and species	Conc.[a]	Conc.[b]	Exposure period (min)	Effects observed	Reference
CFC-11					
Guinea-pig	22-25	125-143	120	tremor, dyspnoea	Nuckolls (1933)
	45-51	257-291	120	tremor, incipient narcosis	
	100	571	50	deep narcosis	Scholz (1961)
	250	1427	30	death (LC_{50})	Caujolle (1964) Paulet (1969)
Mouse	10	57	1440	no clinical signs	Quevauviller et al. (1963)
	100	571	30	death (LC_{50})	Paulet (1969)
Rat	50	285	120	incipient narcosis	Scholz (1961)
	90	514	30	deep narcosis	Lester & Greenberg (1950)
	150	856	30	death (LC_{50})	Paulet (1969)
Hamster	100	571	240	death (LC_{50})	Taylor & Drew (1974)
Cat	100	571	60	death	Scholz (1961)
CFC-12					
Guinea-pig	540	2716	30	initial effect on CNS	Paulet (1969)
	900	4527	30	narcosis, but no mortalities	
Mouse	320	1610	30	initial effect on CNS	Paulet (1969)
Rat	200	1006	30	no effect	Lester & Greenberg (1950)
	300-400	1509-2012	30	tremor	
	500	2515	30	reduced reflexes	
	700-800	3521-4024	30	deep narcosis, no mortality	
	800	4024	360	no mortality	
Monkey, Dog	200	1006	420-480	incoordination	Sayers et al. (1930)
CFC-13					
Guinea-pig	600	2604	120	no effect	Weigand (1971)
Rat	600	2604	120	no effect	Weigand (1971)
CFC-112					
Rat	30	254	40-60	lethal (pulmonary haemorrhage)	Greenberg & Lester (1950)
	5-10	42-85	1080	lethal (pulmonary haemorrhage)	

Table 12 (contd).

Compound and species	Conc.[a]	Conc.[b]	Exposure period (min)	Effects observed	Reference
CFC-113					
Guinea-pig	120	935	60	mortality (LC_{50})	Trockimowitz (1984)
Mouse	90-95	701-740	120	death (LC_{50})	Desoille et al. (1968)
	90	701	120	death (LC_{50})	Trockimowitz (1984)
Rat	52.5	409	240	death (LC_{50})	Trockimowitz (1984)
	110	857	120	death (LC_{50})	
Rabbit	59.5	463	120	death (LC_{50})	Trockimowitz (1984)
CFC-114					
Guinea-pig	20-47	142-334	120	dyspnoea	Nuckolls (1933)
	400	2844	1440	incoordination	Scholz (1961)
Mouse	700	4977	30	no mortality	Paulet (1969)
Rat	300	2133	120	incoordination	Scholz (1961)
	600	4266	120	deep narcosis	
	720	5119	30	no mortality	Paulet (1969)
Rabbit	750	5332	30	no mortality	Paulet (1969)
CFC-115					
Guinea-pig	600	3852	120	no effect	Weigand (1971)
Rat	600	3852	120	no effect	Weigand (1971)

[a] Concentration in parts per thousand.
[b] Concentration in g/m^3.

8.2.1 Inhalation exposure

The results of short-term inhalation studies are summarized in Table 14.

In the case of CFC-11, intermittent exposure to 143 g/m^3 (25 000 ppm) (3.5 h/day, 5 days/week for 4 weeks) did not result in any adverse effects in rats and guinea-pigs (Scholz, 1962). Similarly, no treatment-related effects occurred in rats, guinea-pigs, monkeys, or dogs after intermittent exposure to 58.5 g/m^3 (10 250 ppm) (8 h/day, 5 days/week for 6 weeks) or continuous exposure

Table 13. Acute oral toxicity of various chlorofluoroalkanes to rats[a]

Compound	Approximate lethal dose (mg/kg)
CFC-11	3725[b]
CFC-12	>1000[c]
CFC-112	25 000
CFC-112a	25 000
CFC-113	45 000[d]
CFC-114	>2250[c]

[a] Modified from: Clayton (1966).
[b] From: Slater (1965).
[c] Maximum feasible dose of chlorofluorocarbon dissolved in peanut oil.
[d] LD_{50} = 43 000 mg/kg.

to 57.1 g/m³ (10 000 ppm) for 90 days (Jenkins et al., 1970). Leuschner et al. (1983) exposed groups of dogs to 28.5 g/m³ (5000 ppm) and rats to 57.1 g/m³ (10 000 ppm) for 6 h/day during 90 days and did not find any treatment-related changes. The lowest reported effect level for CFC-11 was 68.5 g/m³ (12 000 ppm), above which pathological changes in the brain, liver, lung, and spleen of all the rats were observed (Clayton, 1966).

The exposure of rats, cats, guinea-pigs, and dogs to CFC-12 at 503 g/m³ (100 000 ppm) (3.5 h/day, 5 days/week for 4 weeks) caused no adverse effects (Scholz, 1962). Fatty infiltration and necrosis in the liver was observed by Prendergast et al. (1967) after continuous (90-day) exposure of guinea-pigs to CFC-12 at 4.02 g/m³ (800 ppm) but not in rats, rabbits, dogs, or monkeys exposed to the same dose regimen. Exposure to CFC-12 for 90 days (6 h/day) of dogs at 25.1 g/m³ (5000 ppm) and rats at 50.3 g/m³ (10 000 ppm) was without any toxic effect (Leuschner et al., 1983).

In most inhalation toxicity studies, CFC-113 caused no adverse effects, even after a 90-day exposure of rats to 155 g/m³ (20 000 ppm) (Trochimowicz, 1984) and dogs to 40 g/m³ (5000 ppm) (Leuschner et al., 1983). However, Clayton (1966) reported effects in rats after 30 exposures

Table 14. Short-term inhalation exposures of various animals to chlorofluorocarbons

Species and number of animals	Exposure g/m³ᵃ	Effects	Reference
CFC-11			
Rat (4)	68 (12 000), 4 h/day, 10 days	pathological changes in lung, liver, brain, and spleen	Clayton (1966)
Rat (5)	143 (25 000), 3.5 h/day, 5 days/week, 4 weeks	no adverse effects	Scholz (1962)
Dog (2)	71 (12 500 ppm, 3.5 h/day, 5 days/week, 4 weeks	no adverse effects	Scholz (1962)
Cat (2) Rat (12) Guinea-pig (2) Rabbit (1)	23 (4000), 6 h/day, 5 days/week 28 exposures	no adverse effects	Clayton (1966)
Rat (15) Guinea-pig (15) Monkey (9) Dog (2)	58 (10 250), 8 h/day, 5 days/week, 6 weeks	no compound-related effects	Jenkins et al. (1970)
Rat (15) Guinea-pig (15) Monkey (9) Dog (2)	5.7 (1008), 24 h/day, 90 days	no compound-related effects	Jenkins et al. (1970)
Dog (6)	28.5 (5000), 6 h/day, 90 days	no adverse effects	Leuschner et al. (1983)
Rat (40)	57.1 (10 000), 6 h/day, 90 days	no adverse effects	Leuschner et al. (1983)

Table 14 (contd).

Species and number of animals	Exposure g/m³ᵃ	Effects	Reference
CFC-12			
Cat (2) Rat (5) Guinea-pig (3) Dog (2)	503 (100 000), 3.5 h/day, 5 days/week	no adverse effects	Scholz (1962)
Guinea-pig (15) Rat (15) Dog (2) Monkey (3)	4.1 (810), 24 h/day, 90 days	fatty infiltration, necrosis in the liver of guinea-pigs, no treatment-related effects in other species	Prendergast et al. (1967)
Guinea-pig (15) Rat (15) Rabbit (3) Dog (2) Monkey (3)	4.1 (800), 8 h/day, 5 days/week	fatty infiltration, necrosis in the liver of guinea-pigs, no treatment-related effects in other species	Prendergast et al. (1967)
Dog Monkey	1006 (200 000), 7-8 h/day	tremor, ataxia, dyspnoea, salivation, lacrimation, no histological changes	Sayers (1930)
Dog (6)	25 (5000), 6 h/day, 90 days	no adverse effects	Leuschner et al. (1983)
Rat (40)	50 (10 000), 6 h/day, 90 days	no adverse effects	Leuschner et al. (1983)

Table 14 (contd).

CFC-112

Rat	8.5 (1000), 18 h/day, 16 days	no toxic effects	Clayton (1967a)
	25 (3000), 4 h/day, 10 days	CNS and respiratory signs	Clayton (1967a)
Rat, mouse, guinea-pig, rabbit	8.5 (1000), 6 h/day, 31 days	slight liver changes	Clayton (1967a)

CFC-113

Rat	16-31 (2000-4000), 6 h/day, 7-14 days	changes in liver enzyme activity, proliferation of smooth endoplasmic reticulum	Vainio et al. (1980)
Rat, mouse, dog, monkey	16 (2000), 24 h/day, 14 days	no adverse effects	Trochimowicz (1984)
Monkey, rat	16 (2000), 24 h/day, 14 days	enlarged thyroids (monkey), increased kidney weight (rat)	Carter et al. (1970)
Mouse, dog	16 (2000), 24 h/day, 14 days	no adverse effects	Carter et al. (1970)
Rat, guinea-pig	195 (25 000), 3.5 h/day, 20 exposures	no adverse effects	Trochimowicz (1984)
Dog	97 (12 500), 3.5 h/day, 29 exposures	no adverse effects	Trochimowicz (1984)
Rat, dog, guinea-pig	40 (5100), 6 h/day, 20 exposures	no adverse effects	Trochimowicz (1984)
Rat	16-22 (2075-2850), 7 h/day, 30 exposures	no adverse effects, reduced rate of body weight gain, pale discoloration of the liver	Clayton (1966)

Table 14 (contd).

Species and number of animals	Exposure g/m³ᵃ	Effects	Reference
Rat	up to 156 (20 000), 6 h/day, 5 days/week, 90 days	no adverse effects	Trochimowicz (1964)
Dog (6)	39 (5000), 6 h/day, 90 days	no adverse effects	Leuschner et al. (1983)
Rat (40)	78 (10 000), 6 h/day, 90 days	no adverse effects	Leuschner et al. (1983)
CFC-114			
Cat, rat, dog, guinea-pig	711 (100 000), 3.5 h/day, 20 exposures	no adverse effects	Scholz (1962)
Guinea-pig (6)	1002 (141 000), 8 h/day, 21 days	occasionally slight fatty degeneration of the liver	Yant et al. (1932)
Guinea-pig (6)	1422 (200 000), 8 h/day, 4 days	occasionally slight fatty degeneration of the liver	Yant et al. (1932)
Dog (3)	1002 (141 000), 8 h/day, 3-21 days	at study start incoordination, tremor, occasionally convulsions; tolerance from day 3-5 onwards	Yant et al. (1932)

Table 14 (contd).

Species (no.)	Exposure concentration[a]	Effects	Reference
Dog (5)	1422 (200 000), 8 h/day, 3-4 days	100% mortality, tremors, convulsions, impaired body weight gain, haematological changes, congestion of all organs	Yant et al. (1932)
Rat (10) Mouse (10)	711 (100 000), 2.5 h/day, 5 days/week, 2 weeks	no adverse effects	Paulet & Desbrousses (1969)
Rat (10) Mouse (10)	1422 (200 000), 2.5 h/day, 5 days/week, 2 weeks	slight impairment of body weight gain increased lymphocyte count, histopathology: alveolar and bronchiolar stasis	Paulet & Desbrousses (1969)
Rat (10) Mouse (10)	71 (10 000), 2.5 h/day, 5 days/week, 2 weeks	no adverse effects	Paulet & Desbrousses (1969)
Dog (6)	36 (5000), 6 h/day, 90 days	no adverse effects	Leuschner et al. (1983)
Rat (40)	71 (10 000), 6 h/days, 90 days	no adverse effects	Leuschner et al. (1983)
CFF-115			
Rat, mouse, dog, rabbit	642 (100 000), 6 h/day, 90 days	no adverse effects	Clayton (1966)

[a] Values in parentheses are exposure concentrations in parts per million.

(each of 7 h) to 40 g/m^3 (5000 ppm) and Vainio et al. (1980) found changes in liver enzyme activities and proliferation of the smooth endoplasmic reticulum after exposure of rats to 15.6-31.2 g/m^3 (2000-4000 ppm) (6 h/day for 7-14 days). Enlarged thyroids in monkeys and increased kidney weights in rats were described after continuous exposure to 15.6 g/m^3 (2000 ppm) for 14 days (Carter et al., 1970), but these effects were minimal and could not be reproduced under the same exposure conditions (Trochimowicz, 1984).

CFC-114 caused no effects in mice, rats, guinea-pigs, cats, or dogs after intermittent exposure to concentrations as high as 711 g/m^3 (100 000 ppm). At higher dose levels (995-1422 g/m^3; 140 000-200 000 ppm) signs of intoxication were noted in guinea-pigs, dogs, rats, and mice (see Table 13). Rats exposed to CFC-112 at 8470 mg/m^3 (1000 ppm) (6 h/day for 31 days) developed slight liver changes. Exposure to the same concentration for 18 h/day during 16 days caused no toxic effects, whereas CNS and respiratory signs occurred at higher concentrations (25 410 mg/m^3; 3000 ppm) (Clayton, 1967). In contrast, CFC-115 at 642 g/m^3 (100 000 ppm) was tolerated by rats, mice, rabbits, and dogs for 90 days (6 h/day) without any toxic effect (Clayton, 1966).

In summary, it can be concluded that short-term inhalation toxicity is low. Toxic effects relate mainly to the central nervous system, the respiratory tract, and the liver.

8.2.2 Oral toxicity

Repeated dose studies of less than 90 days duration have been reported for CFC-12, CFC-112, CFC-112a, CFC-114, and CFC-115 and generally confirm the low toxicity of these chlorofluorocarbons.

In a study by Clayton (1967a), CFC-12 was given orally to rats at a dose level ranging from 160-379 mg/kg per day, as well as to dogs at doses of 84-95 mg/kg per day, for approximately 12 weeks without significant adverse effects related to nutritional, clinical, laboratory, or histopathological indices.

Greenberg & Lester (1950) dosed rats with both CFC-112 and CFC-112a at 2000 mg/kg for 23-33 days with no clinical or histopathological evidence of toxicity. However, when rats were given CFC-112 or CFC-112a at 5000 mg/kg per day for 10 days, Clayton et al. (1966) and Clayton (1967b) found that both chlorofluorocarbons produced tremors, inactivity, initial weight loss, diarrhoea, a slight liver weight increase, and slight, reversible histopathological changes in the liver.

Like CFC-112 and CFC-112a, CFC-114 produced no adverse effects in rats when administered at oral doses of 2000 mg/kg per day for 23-33 days (Quevauviller, 1965). Clayton (1967a) also reported no evidence of toxicity in rats given CFC-114 at 1300 mg/kg per day for 10 days.

Clayton (1966, 1967a) fed CFC-115 to rats at doses of 140-172 mg/kg for 5 days/week for two weeks. No clinical or histopathological effects were seen immediately after the last dose or two weeks later.

8.2.3 Dermal toxicity

McNight & McGraw (1983) investigated the effects on the livers of hairless mice of dermal application of CFC-113. A pad saturated with CFC-113 was applied to an area corresponding to 10% of the body surface for 5 min, twice daily, for 10, 20, or 40 days. No changes occurred in the group exposed for 10 days. Increased vacuolization of liver endoplasmic reticulum was seen after 20 days exposure, which was less pronounced after 40 days, whereas swollen mitochondria were only found after 20 days exposure.

CFC-113 applied to rabbit skin at 5 g/kg per day for 5 days caused gross and histological damage to the skin as well as slight changes in the liver (Clayton, 1966). CFC-11, CFC-12, CFC-113, and CFC-114 at 40% in sesame oil were sprayed onto shaved rabbit skin for 12 exposures with no effect (Scholz, 1962).

When applied to the skin of rabbits at 7.5 g/kg, CFC-112 caused skin erythema but no systematic or histological effects. CFC-112a (11 g/kg) caused histological changes in skin musculature and weight loss (Clayton, 1966).

8.3 Skin and eye irritation; sensitization

Severe local irritation was produced after 5 days by CFC-113 kept occluded in liquified form at 5 g/kg per day on shaved rabbit skin (Waritz, 1971). Quevauviller et al. (1964) and Quevauviller (1965) applied CFC-11, CFC-12, CFC-114, and mixtures of CFC-11 and CFC-12 and of CFC-11 and CFC-22 to the skin, tongue, soft palate, and auditory canal of rats, 1-2 times/day, 5 days/week, for 5-6 weeks. The same compounds were applied once a day, 5 days/week for 1 month to the eye of rabbits. Slight irritation was noted only in the skin of the rats and in the eye of the rabbits. The healing rate of experimental burns on the skin of rabbits, however, was noticeably retarded by all of the compounds.

When applied to the skin of rabbits, CFC-112 at 7.5 g/kg caused erythema and CFC-112a at 11 g/kg caused severe skin irritation, while CFC-113 at 11 g/kg produced only local irritation. CFC-114 did not produce irritation when sprayed directly on the backs of guinea-pigs (Clayton, 1966). CFC-112 produced mild irritation but no sensitization when applied to the skin of guinea-pigs (Clayton et al., 1964).

8.4 Long-term exposures

8.4.1 Inhalation toxicity

In long-term inhalation toxicity studies by Smith & Case (1973), mice and dogs were exposed regularly for brief periods to high levels of mixtures of CFC-11, CFC-12, CFC-114, and CFC-113 (25:49:25:1.1) and CFC-11, CFC-12, CFC-114, and Span 85, an emulsifier, (24.5:50:25:0.5), respectively. In this study 30 female mice were exposed by inhalation 5 days/week at levels of 0 or 970 mg/kg (calculated value) per day for 23 months. No signs of toxicity were observed during the study and there was no evidence of lung tumours after the 23 months of exposure. When adult dogs (three of each sex) were exposed 7 days/week at levels of 0 or 2240 mg/kg (calculated value) per day for 1 year, some signs of toxicity, such as slight depression or drowsiness, were observed in dogs immediately after dosing but lasted only a few minutes. Tissue sections of the lungs did not show signs of toxicity or irritation from inhalation. Also, no changes were observed in haematology, blood chemistry, or urinalysis.

Trochimowicz et al. (1988) performed a 2-year inhalation toxicity and carcinogenicity study of CFC-113 (technical grade, purity 99.89%) on Crl:CD(SD)BR rats (Sprague Dawley-derived). Groups of 100 males and 100 females were exposed to 0, 15.3, 76.6, or 153 g/m^3 (0, 0.2, 1, or 2% in air) 6 h/day, 5 days/week, for up to 2 years. Body weight, appearance and behaviour, and clinical laboratory values (haematology, clinical chemistry, and urinalysis) were monitored regularly. Comprehensive histopathological examinations were performed on rats in the control group and in those exposed to 153 g/m^3. The only findings considered by the authors to be treatment-related were decreases in mean body weight and in body weight gain among females exposed to 76.6 g/m^3 and in both sexes exposed to 153 g/m^3, and a slight, transient increase in serum glucose levels in males exposed to 153 g/m^3.

When rabbits and rats were exposed for a period of 2 years (2 h/day, 5 days per week) to a CFC-113 concentration of about 93 g/m^3 (12 000 ppm), three out of six rats died. In the exposed animals a slight dizziness was observed. Body weight, growth, and haematology values remained unchanged. No compound-related effects on morphology were found which could be attributed to exposure to CFC-113. Also, since three rats died in the control group and no treatment-related signs were seen in the surviving animals exposed to CFC-113, the deaths of the exposed animals were probably not related to the exposure (Desoille et al., 1968).

8.4.2 Oral toxicity

Long-term oral toxicity studies have been carried out on CFC-11 and CFC-12. NCI (1978) completed a bioassay of CFC-11, administered by gavage in corn oil, for possible carcinogenicity in Osborne-Mendel rats and B6C3F$_1$ mice. The assay for carcinogenicity was negative for mice and inconclusive for rats. Time-weighted average doses were 488 and 977 mg/kg per day for male rats, 538 and 1077 mg/kg per day for female rats, and 1962 and 3925 mg/kg per day for male and female mice. All doses were administered

5 days/week. In both male and female rats, a significant (P < 0.001) dose-related acceleration of mortality was noted compared with the vehicle control using the Tarone test. This increase in mortality occurred as early as 4 weeks in the females receiving the higher dose. This early mortality, however, could not be related to changes in body weights, clinical signs, or non-tumour pathology. Low incidences (2-6%) of pleuritis and pericarditis were seen in the treated rats of both sexes at both dose levels but not in the control animals. Chronic murine pneumonia occurred in control and treated animals at incidences of almost 90%. In mice, no statistically significant compound-related effects were noted on weight gain, clinical signs, or non-tumour pathology. In female, but not male, mice, a significant (P=0.009) dose-related increase in mortality was noted compared with the vehicle controls using the Tarone test.

Sherman (1974) conducted long-term oral studies of CFC-12 on rats and dogs. As part of a multi-generation reproductive and chronic toxicity study, groups of 50 male and 50 female Charles River rats of the F_{1a} generation remained in the test for 2 years, with an interim kill at 1 year. Starting at 6 weeks of age, controls, low-dose, and high-dose groups were administered CFC-12 in corn oil or corn oil alone daily by gavage for 6 weeks and 5 times per week thereafter. Over the course of the study, actual daily doses of CFC-12 for low-dose males and females declined from 27 to 11 and 25 to 11 mg/kg per day, respectively, and, for the high-dose males and females, declined from 273 to 130 and 242 to 128 mg/kg per day, respectively. Average doses were 15 mg/kg per day for the low-dose groups and 150 mg/kg per day for the high-dose groups. Body weight gain was depressed in the high-dose groups, particularly among the females, and a slight decline in food efficiency was noted in high-dose females, relative to controls. No overt signs of toxicity were seen, and there were no significant differences between treated and control groups in survival, periodic measurements of haematological, clinical chemistry, and urinalysis values, or in organ weights and histopathological findings. No evidence of carcinogenicity was seen.

In the same study, groups of four male and four female beagle dogs were orally administered CFC-12 (in frozen dog food) at measured doses of 0, 8, or 80 mg/kg per day for 2 years. None of the dogs died or showed signs of toxicity. No significant differences between treated and control groups were found in food consumption, body weight, periodic haematology, clinical chemistry and urine testing, organ weights, or histopathological findings. An adrenal function test (urinary 17-ketosteroid excretion) also revealed no effects. There was no evidence of carcinogenicity (Sherman, 1974).

8.5 Reproduction and developmental toxicity

8.5.1 Reproduction

Effects on reproductive parameters have only been reported for two chlorofluorocarbons, CFC-12 and CFC-113.

In a three-generation, oral gavage study in rats using CFC-12 (in corn oil) at average doses of 15 and 150 mg/kg per day, Sherman (1974) found no adverse effects on reproductive capability as measured by the fertility index (percentage of matings resulting in pregnancy), gestation index (percentage of pregnancies resulting in birth of live litters), viability index (percentage of rats born that survived four days), and lactation index (percentage of rats alive at 4 days that survived to be weaned at 21 days).

In a limited one-generation reproduction study, groups of male and female rats were exposed by inhalation 6 h/day, 5 days per week, for 10 weeks (males) or 3 weeks (females) to CFC-113 at either 39 or 97 g/m^3 (5000 or 12 500 ppm). Each male rat was then paired with two females for 2 weeks during which time exposure was 6 h/day, 7 days/week. Females that showed positive signs of mating continued to be exposed 6 h/day until day 20 of gestation when they were allowed to give birth. The development of their offspring was followed for up to 4 weeks. There were no adverse effects on any of the standard reproductive indices (US EPA, 1983).

8.5.2 Developmental toxicity

Of the chlorofluorocarbons reviewed in this document, developmental toxicity studies have been reported for CFC-11, CFC-12, and CFC-113.

Effects on Experimental Animals and In Vitro Test Systems

In a study of a mixture (10% CFC-11 and 90% CFC-12), groups of rats and rabbits were exposed by inhalation on days 4-16 (rats) or days 5-20 (rabbits) of gestation for 2 h/day at a concentration of 1558 g/m^3 (200 000 ppm). No evidence of embryotoxicity, fetotoxicity, or teratogenicity was seen when rats and rabbits were sacrificed at 20 days (rats) or 30 days (rabbits) of gestation. In addition, offspring of the dams that were allowed to deliver naturally showed no evidence of toxicity relative to survival or growth (US EPA, 1983).

In another study, groups of 25 to 27 pregnant Charles River rats were given CFC-12 in corn oil by gavage at doses of 16.6 or 179 mg/kg per day on days 6-15 of gestation. Neither dose induced any evidence of embryotoxicity or teratogenicity (Sherman, 1974).

Three unpublished studies on rats and rabbits were carried out using CFC-113 and were reviewed by the US EPA (1983). Groups of 24 pregnant rats were exposed 6 h/day on days 6-15 of gestation to either 39, 97, or 195 g/m^3 (5000, 12 500, or 25 000 ppm in air). Some evidence of maternal toxicity (reduced weight gain, decreased food intake) was seen at the highest exposure level. However, there was no evidence of embryotoxicity, fetotoxicity, or teratogenicity at any exposure level.

Two studies using rabbits were judged to be inadequate by the US EPA (1983) because of the small number of dams and fetuses evaluated, the limited exposure time each day, and excessive maternal toxicity. In the first study (Hazleton Laboratories, 1967a), groups of 12 rabbits were exposed to CFC-113 at either 15.6 or 156 g/m^3 (2000 or 20 000 ppm in air) for 2 h/day on days 8-16 of gestation. Signs of maternal toxicity were seen at the highest exposure level but there was no evidence of embryotoxicity or teratogenicity attributable to CFC-113.

In the second rabbit study (Hazleton Laboratories, 1967b), groups of eight rabbits were given CFC-113 at either 1000 or 5000 mg/kg per day by gavage on days 8-11 of gestation. No unusual skeletal or visceral abnormalities were observed at either dose level, but low pregnancy rates and fetal deaths were seen in control and test groups.

In conclusion, none of the three chlorofluorocarbons tested (CFC-11, CFC-12, and CFC-113) show any evidence of reproductive or developmental toxicity. There is no significant information on this subject for any of the other chlorofluorocarbons evaluated in this monograph.

8.6 Mutagenicity and related end-points

The mutagenic potential of the chlorofluorocarbons reviewed in this monograph has been evaluated, primarily using the *Salmonella* assay (Uehleke et al., 1977; Longstaff et al., 1984) with negative results. Negative results were also obtained for CFC-11, CFC-12, and CFC-115 in a cell transformation assay (Longstaff et al., 1984) and for CFC-11 and CFC-12 in a mammalian cell mutagenicity test (Krahn et al., 1982). CFC-12 was also tested in a plant assay using *Tradescantia*, and found to be negative (Van't Hof & Schairer, 1982).

A dominant lethal assay was performed as part of a reproduction study on rats using CFC-12 at doses of 15 and 150 mg/kg per day (gavage) for several weeks (Sherman, 1974). Another dominant lethal assay was performed with CFC-112 and CFC-113 on mice after single intraperitoneal injections of 200 and 1000 mg/kg (Epstein et al., 1972). Negative results were obtained in both of these *in vivo* assays. The mutagenicity studies are summarized in Table 15.

8.7 Carcinogenicity

Long-term oral carcinogenicity studies of CFC-11 (NCI, 1978) and CFC-12 (Sherman, 1974) gave negative results. When administered by gavage to groups of 50 male and 50 female $B6C3F_1$ mice (see section 8.4.2), CFC-11 at 1962 or 3952 mg/kg per day, 5 days/week, for 78 weeks, followed by 13 weeks of observation, produced no evidence of carcinogenicity (NCI, 1978). Gavage administration of CFC-11 to groups of 50 male and 50 female Osborne-Mendel rats in the same study also produced no evidence of carcinogenicity, but the results were considered to be inconclusive by the NCI (1978) because the numbers of rats surviving long enough to be at risk from late-developing tumours were insufficient. In this study, time-weighted average doses

Table 15 Mutagenic Assays of Chlorofluorocarbons

Compound (code)	Method or type of assay (exposures)	Organisms	Strain	Use of metabolic activating system	Results	Reference
CFC-11	Reverse mutation[a] (L)	S. typhimurium	TA1535	yes	negative	Uehleke et al. (1977)
CFC-11	Reverse mutation[a] (L)	S. typhimurium	TA1538	yes	negative	Uehleke et al. (1977)
CFC-11	Reverse mutation (G)	S. typhimurium	TA100	no/yes	negative[d]	Longstaff et al. (1984)
			TA1535	no/yes	negative	
CFC-11	Forward mutation (G) HGPRT assay	CHO cells	NA	no/yes	negative	Krahn et al. (1982)
CFC-11	Cell transformation (G)	BHK21 cells	NA	yes	negative	Longstaff et al. (1984)
CFC-12	Reverse mutation (G)	S. typhimurium	TA100	no/yes	negative[d]	Longstaff et al. (1984)
			TA1535	no/yes	negative	
CFC-12	Forward mutation (G) HGPRT assay	CHO cells	NA	no/yes	negative	Krahn et al. (1982)
CFC-12	Cell transformation (G)	BHK21 cells	NA	yes	negative	Longstaff et al. (1984)
CFC-12	Forward mutation (G) Blue locus	Trandescantia hybrid	Clone 4430[c]	NA	negative	Van't Hof & Schairer, (1982)
CFC-12	Dominant lethal mutation (g.i.)	rats[b]	Charles River CD	NA	negative	Sherman (1974)
				NA	negative	

Table 15 (contd).

CFC-13	Reverse mutation (G)	S. typhimurium	TA100 TA1535	no/yes no/yes	negative[d] negative	Longstaff et al. (1984)
CFC-112	Dominant lethal mutation (i.p.)	Swiss mice[b]	ICR/Ha	NA	negative	Epstein et al. (1972)
CFC-113	Reverse mutation (G)	S. typhimurium	TA100 TA1535	no/yes no/yes	negative[d] negative	Longstaff et al. (1984)
CFC-113	Dominant lethal mutation (i.p.)	Swiss mice[b]	ICR/Ha	NA	negative	Epstein et al. (1972)
CFC-114	Reverse mutation (G)	S. typhimurium	TA1535[c]	no/yes	negative[d]	Longstaff et al. (1984)
CFC-115	Reverse mutation (G)	S. typhimurium	TA1535	no/yes	negative[d]	Longstaff et al. (1984)
CFC-115	Cell tranformation (G)	BHK21 cells	NA	yes	negative	Longstaff et al. (1984)

[a] Ames assay performed with pre-incubation for 60 min in closed vials under nitrogen gas.
[b] See text for dose information.
[c] Heterozygous for blue locus.
[d] Longstaff et al. (1984) stated that all the fluorocarbons tested gave negative results in S. typhimurium TA1538 and TA98 in the absence or presence of a metabolic activating system.

NA = not applicable; G = gas; L = liquid; g.l. = gastric intubation; i.p. = intraperitoneal; BHK21 = permanent cell line of baby hamster kidney fibroblasts; CHO = Chinese hamster ovary; HGPRT = hypoxanthine guanine phosphoribosyl transferase.

of CFC-11 (488 and 977 mg/kg per day for male rats and 538 and 1077 mg/kg per day for female rats) were administered 5 days/week for 78 weeks, followed by 28-33 weeks of observation.

CFC-12 administered by gavage at doses of 15 or 150 mg/kg per day for 2 years to groups of 50 male and 50 female Charles River rats of the F_{1a} generation in a multi-generation study (see section 8.4.2) produced no evidence of carcinogenicity (Sherman, 1974).

Data on the carcinogenicity of inhaled CFC-11 and CFC-12 by Sprague-Dawley rats and Swiss mice have been reported by Maltoni et al. (1988). When administered to groups of 90 male and 90 female rats and 60 male and 60 female mice at concentrations of 1000 or 5000 ppm (57 or 285 g CFC-11/m^3; 49 or 247 g CFC-12/m^3), 4 h/day, 5 days/week, neither compound was found to have induced statistically significant differences in the incidence of total benign or malignant tumours when compared with groups of unexposed rats or mice. The authors also stated that the incidence of all tumours and of some particularly frequently occurring spontaneous tumours in mice showed a tendency to increase in animals exposed to CFC-11 and CFC-12 and that the increased incidence was usually observed in one sex and was not always dose related, possibly due to a longer survival of the treated mice compared with controls.

Trochimowicz et al. (1988) conducted a 2-year inhalation toxicity and carcinogenicity study of CFC-113 in Crl:CD(SD)BR rats. In this study, groups of 100 males and 100 females were exposed to CFC-113 (technical grade, purity 99.89%) at 0, 152, 760, or 1520 g/m^3 (0, 2000, 10 000, or 20 000 ppm), 6 h/day, 5 days/week, for up to 2 years. The toxic effects, which were minimal, are reviewed in section 8.4.1. Primary nasal tumours were found in one male rat exposed to 1520 g/m^3 and in three male rats and one female rat exposed to 760 g/m^3. The five nasal tumours were classified as an adenoma, an undifferentiated sarcoma, an early carcinoma, a carcinoid-like neoplasm, and a papilloma. The authors noted that spontaneous nasal tumours were rare in the control rats. Female rats exposed to 1520 g/m^3 showed a statistically significant increase, relative to control females, in the incidence of pancreatic islet cell adenomas. The incidence of this tumour in the females of the highest dose group was 5.8%. Because the nasal tumours found in the treated rats were of various morphological types and the incidences were not dose related, the authors concluded that the occurrence of this tumour was not related to CFC-113 exposure. However, as indicated in section 8.3, CFC-113 is a local irritant at high concentrations. Thus the neoplasms found in the nasal cavity may be the consequence of the local irritation caused by CFC-113 in the peculiar anatomical configuration of this cavity in the rat. In addition, because the incidence of pancreatic islet cell adenomas in the females exposed to 1520 g/m^3 was "within the anticipated incidence of this finding among untreated Crl:CD(SD)BR rats," the increased incidence in this group, relative to matched controls, was not considered to be related to CFC-113 exposure.

CFC-112 and CFC-113, at a dose of 0.1 ml 10% (v/v) solution injected subcutaneously into the neck of neonatal mice, were not carcinogenic. However, when injected in conjunction with a 5% (v/v) solution of piperonyl butoxide, hepatomas were induced in male mice. This was particularly marked with CFC-113. The apparent synergistic hepatocarcinogenicity of these chlorofluorocarbons with piperonyl butoxide cannot be explained at present. The investigators speculated that piperonyl butoxide may interfere with the metabolism of chlorofluorocarbons (Epstein et al., 1967a). The significance of this effect is difficult to interpret because of the lack of follow-up studies in other species (Tomatis et al., 1973) and with other chlorofluorocarbons.

8.8 Special studies - cardiopulmonary effects

8.8.1 Cardiac sensitization in response to exogenous adrenaline-induced arrhythmia

A variety of hydrocarbons, with and without halogen substitution, have long been known to sensitize the heart to adrenaline-induced arrhythmias including ventricular fibrillation (Hermann & Vial, 1935; Garb & Chenoweth, 1948; Hays, 1972; Reinhardt et al., 1973). At various concentrations, chlorofluorocarbons have been shown to produce this effect. Because this arrhythmogenic action may be related to a variety of human health hazards, a great deal of research has been stimulated in this area, focused primarily on determining the minimum concentration of chlorofluorocarbons and adrenaline required to produce arrhythmias in various mammals. The reader is referred to Zakhair & Aviado (1982) for a review of the literature on this subject.

Reinhardt et al. (1971b) exposed dogs to varying concentrations of CFC-12 for periods of 0.5-10 min and found that a minimum concentration of CFC-12 in air of 250 g/m^3 (5%) was necessary to sensitize the heart to an intravenous dose of adrenaline (8 µg/kg body weight) and that increasing the period of exposure to lower concentrations did not result in arrhythmias. However, increasing the chlorofluorocarbon concentration resulted in a reduction of the sensitizing concentration threshold of exogenous adrenaline. In dogs exposed to normal oxygenation and CFC-12 at 500 g/m^3 (10%) (which resulted in 75 µg CFC-12/ml in arterial blood), the exogenous adrenaline concentration related to arrhythmia was 3 µg/kg per min, while it was only 2.5 µg/kg per min at a CFC-12 concentration of 1000 g/m^3 (20%) (resulting in 155 µg CFC-12/ml in arterial blood) (Lessard et al., 1977b). For the same CFC-12 concentration (1000 g/m^3, 20%, resulting in 155 µg/ml in arterial blood), arrhythmias occurred in rabbits only at the concentration of 8 µg/kg per min of exogenous adrenaline, the arrhythmogenic threshold depending also on the animal species (Lessard et al., 1977a). For the most part, the comparative arrhythmogenic potencies of chlorofluorocarbons are similar to those noted in standard inhalation studies: as fluorination increases within a homologous series, toxicity tends to decrease. Thus, for the chlorofluoromethanes, the arrhythmogenic potency of CFC-11 seems to be greater than that of CFC-12. A similar pattern is seen in the fully halogenated ethanes (CFC-113 > CCF-114 > CFC-115) (Reinhardt et al., 1971b, 1973; Clark & Tinston, 1972a,b; Wills, 1972).

That a critical blood level of chlorofluorocarbon is needed to cause sensitization indicates that differences among the chlorofluorocarbons may primarily reflect differences in absorption characteristics rather than any toxic mechanisms on the molecular level (Jack, 1971;

Taylor et al., 1971; Clark & Tinston, 1972a; Azar et al., 1973). The similarities in lowest venous blood concentrations associated with cardiac sensitization in these various studies suggest that these compounds act in a similar and perhaps non-specific manner in causing arrhythmias. This type of speculation is at least circumstantially supported by the basic similarities in cardiac effects caused by these and other halo-substituted hydrocarbons.

8.8.2 Cardiac sensitization and asphyxia-induced arrhythmia

Studies by Taylor & Harris (1970a) indicated that inhaled chlorofluorocarbons are toxic to the hearts of mice, as shown by the rapid onset of sinus bradycardia and atrioventricular (AV) block induced by a degree of partial asphyxia sufficient to cause tachycardia in unexposed mice. However, four other groups of investigators (Azar et al., 1971; Jack, 1971; Egle et al., 1972; McLure, 1972) failed to confirm these findings. They found that the effects caused by chlorofluorocarbons do not vary significantly from those caused by nitrogen asphyxia controls. These authors concluded that the bradycardia and AV block were actually due to asphyxia, not to chlorofluorocarbon exposure. Although the preceding controversy was never resolved, it is important to note that no reports have been found that confirm the work of Taylor & Harris (1970a).

The mechanism by which arrhythmia is related to the severity of asphyxia is at present unknown. However, it may be related to the degree of adrenaline stimulation or myocardial depression associated with a given level of asphyxia (Zakhair & Aviado, 1982). This mechanism is also probably related to the nature of the chlorofluorocarbon, since induced changes in heart rate depend on chlorofluorocarbon concentrations and the animal species. For example, inhalation of CFC-12 (concentrations of CFC-12 in gas mixture are given in brackets; animals anaesthetized unless stated otherwise) has been reported to cause:

- bradycardia in mice (1006 and 2012 g/m^3, i.e. 20 and 40%) (Aviado & Belej, 1974), rats (1006 g/m^3, 20%) (Doherty & Aviado, 1975), and dogs (concentration not given) (Flowers & Horan, 1972);

- no change in cardiac rate in monkeys (251 and 503 g/m³, i.e. 5 and 10%) (Belej et al., 1974), rats (from 0-2515 g/m³, i.e. 0 to 50%) (Friedman et al., 1973), cats (1257 g/m³, 25%) (Harris et al., 1971), or rats (503, 1006, and 2012 g/m³, i.e. 10, 20, and 40%) (Watanabe & Aviado, 1975);

- tachycardia in monkeys (251-503 g/m³, i.e. 5-10%) (Aviado & Smith) and in unanaesthetized rats (503, 1006, and 2012 g/m³, i.e. 10, 20 and 40%) (Watanabe & Aviado, 1975).

According to Lessard et al. (1977a), CFC-12 causes tachycardia in anaesthetized rabbits at concentrations of 1006 g/m³ (20%) or more and in anaesthetized dogs at concentrations of 503 g/m³ (10%) or more (Lessard et al., 1977b). After surgical removal of baroreceptors in rabbits and dogs, CFC-12 at any concentration causes bradycardia. Thus, the tachycardia is due to the baroreflex secondary to the fall in arterial blood pressure (Lessard et al., 1978).

Hypoxia is also known to potentiate the myocardial depression due to CFC-12 in isolated papillary muscle of rats (Kilen & Harris, 1972, 1976) and in the heart of anaesthetized rabbits (Taylor & Drew, 1974).

Hypoxia associated with high concentrations of CFC-12 (4778 g/m³, i.e. 95%, in gas mixture or 1475 µg/ml in arterial blood) is needed to cause fatal arrhythmia in anaesthetized dogs (Flowers & Horan, 1975).

8.8.3 Arrhythmia not associated with asphyxia or adrenaline

Some chlorofluorocarbons have been found to affect cardiac function under conditions of adequate oxygenation or in the absence of elevated adrenaline levels. Arrhythmia, in the absence of hypoxaemia or hypercarbia, has been demonstrated in dogs (CFC-12, 3974 g/m³, i.e. 79%, resulting in 725 µg/ml in arterial blood, Lessard et al., 1978; CFC-12 concentration not mentioned, Flowers & Horan, 1972), monkeys (CFC-12, 1509 g/m³, 30% + CFC-114, 453 g/m³ (9%): Taylor et al., 1971) and rabbits (79% CFC-12 resulting in 855 g/ml in arterial blood: Lessard et al., 1978). However, Lessard et al. did not obtain arrhythmia in either dogs or rabbits, exposed to the same concentrations of CFC-12, after surgical removal of baroreceptors, showing that these arrhythmias were related to reflex endogenous adrenaline delivery (Lessard et al., 1978).

8.9 Mechanisms of toxicity - mode of action

Studies concerning cardiac arrhythmias caused by chlorofluorocarbons at high concentrations (see section 8.8) may be interpreted in two ways (Taylor et al., 1971). Firstly, the chlorofluorocarbon gases may be exerting some direct effect on the myocardium. Secondly, they may have sensitized the ventricular myocardium to endogenous catecholamines (see Aviado & Belej, 1974). This latter interpretation is consistent with the blocking of arrhythmias by propranolol. The two interpretations are supported by the more recent results of Lessard et al. (1980), Lessard & Paulet (1985), and Lessard & Paulet (1986). On the basis of an electrophysiological analysis of the action of CFC-12 on different types of cardiac cells from rats and sheep, these authors conclude that:

- the cardiac depression observed during inhalation of CFC-12 (and many other volatile liposoluble compounds) is the consequence of a non-specific impairment of the membrane properties and notably the inhibition of trans-membrane ionic currents;
- CFC-12 action on ionic currents is variable: at high concentrations, depending on the type of cardiac cell, it can oppose or favour the action of adrenaline, giving rise to many factors that lead to arrhythmia.

In a review of the toxicity of chlorofluorocarbons, Aviado (1978) proposed a combination of mechanisms for the cardiopulmonary effects of fluorocarbons. Disintegration of normal alveolar surfactant was implicated in the death of an adolescent after abuse of a chlorofluorocarbon-propelled aerosol (Fagan et al., 1977).

Several investigators have attempted to determine if chlorofluorocarbons affect oxidative phosphorylation. Griffin et al. (1972) showed that CFC-12 and CFC-114 do not markedly affect oxygen consumption or oxidative phosphorylation in mitochondria isolated from the liver, lung, brain, heart, or kidney of rats exposed to about 7.5% chlorofluorocarbons prior to mitochondrial isolation. Further *in vitro* studies were conducted with liver and heart mitochondria in which measurements were taken during exposure of the mitochondria to CFC-12 at 990 g/m^3 (20%) (time of exposure not specified). No effects on either oxidation or phosphorylation were noted (Griffin et al., 1972).

9. EFFECTS ON HUMANS

9.1 Controlled studies with volunteers

The studies reported by Stewart et al. (1978) are among the more comprehensive attempts to characterize the effects of CFC-11 and CFC-12 on human volunteers under controlled conditions. A number of biological end-points, including clinical haematology and chemistry, ECG, EEG, pulmonary function, neurological parameters, and cognitive tests were monitored. Single exposures to CFC-11 or CFC-12 at concentrations of 0.025% (CFC-11, 1.4 g/m^3; CFC-12, 1.2 g/m^3), 0.05% (CFC-11, 2.8 g/m^3; CFC-12, 2.5 g/m^3), and 0.1% (CFC-11, 5.6 g/m^3; CFC-12, 5 g/m^3), for 1 min to 8 h, induced no observable effects. There was a statistically-significant decrease in cognitive test performance in subjects exposed to CFC-11 at 5.6 g/m^3, 8 h/day, 5 days/week, for 2-4 weeks, but not in subjects exposed to CFC-12 at 5 g/m^3. Although the authors regarded this effect as "spurious", the similar effects observed on acute exposures to CFC-12 (Kehoe, 1943; Azar et al., 1972) and CFC-113 (Stopps & McLaughlin, 1967), as well as the behavioural effects of CFC-21 on baboons (Geller et al., 1977), suggest that psychomotor impairment may be a general effect of fluorocarbons, which precedes signs of definite toxic effects.

Kehoe (1943) exposed one subject to CFC-12 at concentrations of 198 g/m^3 (4%), 297 g/m^3 (6%), 347 g/m^3 (7%), and 545 g/m^3 (11%) for periods of 80, 80, 35, and 11 min, respectively. A second subject was exposed to 198 g/m^3 for 14 min, immediately followed by 99 g/m^3 (2%) for 66 min. At 198 g/m^3, the subject experienced a tingling sensation, humming in the ears, and apprehension. EEG changes were noted as well as slurred speech and decreased performance in psychological tests. In the subject exposed to higher concentrations, these signs and symptoms became more pronounced with increases in concentration. An exposure to 545 g/m^3 for 11 min caused a significant degree of cardiac arrhythmia, followed by a decrease in consciousness with amnesia after 10 min. At a concentration of 50 g/m^3 (1%) for 150 min, Azar et al. (1972)

noted a 7% decrease in psychomotor test scores, but no effect at 5 g/m^3 (0.1%) over the same period. Valić et al. (1977) exposed 10 subjects to CFC-11, CFC-12, CFC-114, two mixtures of CFC-11 and CFC-12, and a mixture of CFC-12 and CFC-114 (breathing concentrations between 16 and 150 g/m^3) for 15, 45, or 60 seconds, and found significant acute reduction of ventilatory lung capacity (FEF50, FEF25) on exposure to each chlorofluorocarbon, as well as bradycardia and increased variability in heart rate in seven subjects, negative T-waves in two subjects (one was exposed to CFC-11 and CFC-12), and atrioventricular block in 1 subject (CFC-114). Mixtures exerted stronger respiratory effects than individual chlorofluorocarbon at the same level of exposure.

In another study, 11 subjects (7 being maintenance technicians of large cooling and refrigerating systems) were exposed for 130 min to CFC-12 (weighted exposure 0.46, 49.9, and 87.7 g/m^3), HCFC-22 (0.71 and 18.9 g/m^3), and CFC-502 (a mixture of 4 g CFC-115/m^3 and 1.4 g CFC-22/m^3 or of 23.4 g CFC-115/m^3 and 10.5 g CFC-22/m^3). This led to acute reduction of ventilatory lung capacity only at the two highest CFC-12 concentrations, under which conditions a significant decrease in the heart frequency was also observed (Valić et al., 1982).

CFC-113 has been tested on human subjects by Stopps & McLaughlin (1967) and Reinhardt et al. (1971a). Psychomotor performance was evaluated with exposures to 12 g/m^3 (0.15%), 19 g/m^3 (0.25%), 27 g/m^3 (0.35%), and 35 g/m^3 (0.45%) for 165 min (Stopps & McLaughlin, 1967). At the lowest level, no effect was noted, but at 19 g/m^3 there was difficulty in concentrating and some decrease in test scores. These effects were more pronounced at 27 g/m^3, and at 35 g/m^3 performance in various tasks was decreased by between 10 and 30%. These decreases coincided with sensations of "heaviness" in the head, drowsiness, and a slight loss of orientation after shaking the head from left to right. Reinhardt et al. (1971a) exposed four human subjects to CFC-113 at concentrations of 8 g/m^3 (0.1%) and 4 g/m^3 (0.05%) for 180-min periods in the morning and afternoon on 5 days. No decreases in psychomotor ability were noted. No abnormal findings were noted during postexposure physical examination, haematological and blood chemistry tests (conducted 3 days after final exposure), and steady-state measurements of diffusing capacity of lungs and fractional uptake of carbon monoxide.

Skuric et al. (1975) exposed 17 volunteers for 2 min to 8 household sprays containing CFC-11 (19-54 g/m^3), CFC-12 (12-106 g/m^3), or both. They observed significant reductions in ventilatory lung capacity in each case (relative reductions: FEV1, 3.3-7.4%; FEF50, 5.3-11.2%; FEF25, 12.1-20.9%). However, the reductions on exposure to sprays were greater than those measured in separate exposures to chlorofluorocarbons only (at corresponding breathing concentrations), suggesting that other spray components were mainly responsible for the changes in ventilatory function. Similar findings of acute ventilatory capacity reductions on exposure to hair-sprays were described by Zuskin et al. (1974, 1981) and Swift et al. (1979), but exposure levels were not given in these papers.

Graff-Lonnevig (1979) reported a study of the effects of chlorofluorocarbons on bronchiolar tone in asthmatic children. Forced expiratory volume (FEV), a measure of bronchial tone, was measured in 18 children with a history of asthma, before and after inhaling aerosols of the β_2-receptor agonist, fenoterol, or a mixture of CFC-11, CFC-12, and CFC-114, and in the absence of treatment. The levels of exposure were not reported. Exposure to the chlorofluorocarbon mixture significantly reduced FEV for 2 h, relative to "no treatment", and for 8 h relative to exposure to fenoterol (containing CFC-11 and CFC-12). The results suggest that chlorofluorocarbons can decrease bronchial tone in asthmatic patients, but that this effect is transient and of a sufficiently small magnitude to be superseded by the dilating effects of fenoterol when both fenoterol and chlorofluorocarbon propellants are inhaled together.

Van Ketel (1976) reported allergic contact eczema in patch tests performed on three patients that had a prior history of skin reactions to deodorant sprays. All three patients showed strong positive reactions to 11 deodorant sprays and mild to strong reactions to CFC-11. One patient showed a mild reaction to CFC-12. Fifteen controls (without prior history of allergy to deodorants) showed no response to either CFC-11 or CFC-12. These results suggest that individuals may become sensitized to certain chlorofluorocarbons applied repeatedly to the skin surface.

9.2 Occupational exposure

Two studies (Imbus & Adkins, 1972; NIOSH, 1980) suggest that frequent occupational exposures to CFC-113 do not pose a serious health hazard. No adverse effects occurred at levels as high as 36.7 g/m^3 (0.478%) and averaging 5.4 g/m^3 (0.07%).

Significant acute reductions in the ventilatory lung capacity during a work shift of hairdressers using chlorofluorocarbon-containing hair-sprays were observed in several studies (Zuskin & Bouhuys, 1974; Valić et al., 1974; and Zuskin et al., 1974).

Several cases of accidental death attributed to occupational exposure to chlorofluorocarbons have been reported. May & Blotzer (1984) described three cases of death from exposure to CFC-113. In each case, the individuals succumbed to high concentrations of CFC-113 vapour. A level of 997 g/m^3 (128 000 ppm) was estimated in one case and, in another, death occurred within 15 min after exposure to an estimated 47-288 g/m^3 (6000-37 000 ppm).

Cases of neurological effects attributed to occupational exposure to chlorofluorocarbons have been reported. Raffi & Violante (1981) described a case of neuropathy in a laundry worker exposed to tetrachloroethylene for 6 years and to undetermined levels of CFC-113 for 7 years prior to reporting symptoms. Both dermal and inhalation exposure probably occurred. The individual experienced pain, paraesthesia and weakness of the legs, and decreased motor nerve conduction velocity. She was instructed to avoid contact with CFC-113 and to rest; her clinical condition improved without treatment in one month.

A similar case concerning a refrigerator repair worker was reported by Campbell et al. (1986). Symptoms included pain, paraesthesia, and weakness in the legs, and low nerve conduction velocities. The case was subsequently followed up with a study of 27 refrigerator repair workers. The refrigerator workers and a reference group of

14 pipe fitters and insulators were given a medical examination that included measurements of nerve conduction velocities. The refrigerator workers reported a significantly elevated incidence of "lightheadedness" and palpitations, but no differences in nerve conduction velocities were observed.

In another study, 89 workers were examined during their work with refrigerant equipment. The refrigerants used were mainly CFC-12 (in 56% of the cases) and HCFC-22 (32%), the rest being CFC-11, CFC-500 (a mixture of CFC-12 and HCFC 152a), and CFC-502 (a mixture of CFC-115 and HCFC 22). The mean exposure time was 10 min. Chlorofluorocarbon concentrations in the breathing zone were measured for each person individually. The levels exceeded 750 ppm at least once (as one minute mean values) for 60 of the 89 individuals. Cardiac arrhythmias were registered before, during, and after the exposure by means of a portable ECG instrument connected to a tape recorder. No statistically significant difference was found between exposed and non-exposed periods, nor was there any dose-related trend for different individuals when grouped into different exposure groups. In this study, possible effects on the central nervous system were also studied by means of simple reaction time measurements before and after the exposure. No impairment was seen (Edling & Ohlson, 1988).

Szmidt et al. (1981) investigated death rates among 539 workers exposed occupationally in constructing and repairing refrigeration equipment. The chlorofluorocarbons used were CFC-12, HCFC-22, and CFC-502 (a mixture of CFC-115 and HCFC-22). No increase in total deaths (18 cases) was seen among those employed more than 6 months, compared to the expected number (26 cases), nor was there any statistically significant increase in total tumour deaths or deaths caused by lung cancer or cardiovascular diseases. When the study was restricted to those exposed for more than 3 or 10 years, still no significant increases were seen. No data on exposure levels were given.

Thomas (1965) reported that workers who spilled a large volume of CFC-11 were exposed to high concentrations and developed narcotic effects. In one case, unconsciousness occurred, and in another, potentiation of the endogenous adrenaline effect and tachycardia.

Effects on Humans

Yonemitsu et al. (1983) reported an industrial accidental death due to exposure to CFC-113 used as a solvent for cleaning a washer tub filter. A large quantity of liquid CFC-113 was found at the bottom of the washer tub in which the filter had been cleaned. The concentration of the CFC-113 was 935-1091 g/m^3 (12-14%). Death was attributed to inhalation of the highly concentrated CFC-113 vapour in the washer room.

The US National Institute for Occupational Safety and Health (May & Blotzer, 1984) reported the deaths of 12 workers due to asphyxiation or cardiac arrhythmia resulting from excessive occupational exposure to CFC-113 while working in confined spaces or areas with insufficient ventilation. Uncontrolled use of CFC-113 as a solvent was reported as the primary cause for these deaths. This report contains recommendations for controlling exposure to CFC-113.

9.3 Non-occupational exposures

Only two cases of accidental ingestion of chlorofluorocarbons have been reported. Clayton (1966) reported that 1 litre of CFC-113 was accidentally released into the stomach of an anaesthetized patient, causing transient cyanosis. For the next 3 days, the patient experienced severe rectal irritation and diarrhoea.

A study by Marier et al. (1973) involved a group of 20 housewives who were given 13 household products containing chlorofluorocarbon propellant (CFC-11, CFC-12, and CFC-114) to be used during 4 weeks in conformity with a prescribed protocol. The only effect noted was an increase in lactate dehydrogenase (LDH) during the exposure period, which nonetheless remained within normal limits.

Exposure to CFC-11 and CFC-12 has been associated with the abusive inhalation of aerosols. Bass (1970) concluded that deaths associated with abusive aerosol inhalation were probably caused by cardiac arrhythmia, possibly aggravated by elevated levels of catecholamines due to stress or by moderate hypercapnia. This deduction was subsequently supported by a variety of investigators who found that many chlorofluorocarbons can sensitize the hearts of various mammals to adrenaline, resulting in serious arrhythmias or death (section 8.8).

There are three possible explanations for deaths among asthmatics after using anti-asthmatic drug aerosol formulations containing chlorofluorocarbons as propellant: ineffectiveness of the anti-asthmatic drug; drug overdose; toxicity of CFC-11 and other chlorofluorocarbon propellants. However, the amount of chlorofluorocarbon contained in the inspired aerosols when used correctly for the intended purpose is small compared to that used in the animal experiments of Taylor & Harris (1970b), or under conditions of severe abuse. This is supported by the animal experiments on the dose-dependent relationship between cardiotoxic effects of chlorofluorocarbons and adrenaline described in section 8.8.1.

9.4 Health effects associated with stratospheric ozone depletion

There is undisputed evidence that the atmospheric concentrations of chlorofluorocarbons deplete ozone in the stratosphere. A reduction in ozone concentration will result in increased transmission of solar ultraviolet radiation through the stratosphere. Many significant adverse effects of such an increase in exposure to this radiation have been identified.

The information summarized below indicates that stratospheric ozone depletion has the potential to exert very substantial effects on human health.

9.4.1 Skin cancer effects

One of the most well-defined human health effects resulting from stratospheric ozone depletion is an increase in the frequency of skin cancer expected as a result of even small increases in UV-B radiation (280-320 nm) reaching the earth's surface (US EPA, 1987a; Kripke, 1989; van der Leun, 1988).

The most definitive evidence links the incidence of non-melanoma basal and squamous cell carcinomas to UV-B radiation. These carcinomas occur most frequently on the sun-exposed skin of light-skinned Caucasian people and their incidences increase with age. The geographical distribution data suggest a relationship to cumulative lifetime exposure to sunlight (Urbach, 1969). Increases in skin cancers during the past few decades in the USA are probably partly due to increasing exposure to natural and artificial sources of UV-B radiation and partly to greater longevity allowing for the appearance of such cancers after long latencies (several decades). Although this long latency makes it unlikely that the increases already observed in skin cancer rates are due to the small decreases in stratospheric ozone observed during the past decade, there is extensive evidence for predictions of further increases in basal and squamous cell skin cancer rates if stratospheric ozone depletion continues. Such depletion will result in greater increases in UV-B radiation near to 280 nm rather than in the upper end of the affected wavelength range (320 nm). Skin cancers are mostly induced by UV radiation at around 300 nm, as demonstrated by experimental animal studies. Tumour incidence is a function of the dose received regularly and the period over which doses occur (frequently corresponding to age). There is no evidence of any threshold. A greater-than-linear growth of cancer incidence rate can be predicted due to several factors including optical amplification and bioamplification (van der Leun & Daniels, 1975).

The optical amplification factor indicates the percentage increase in the carcinogenically effective radiation caused by a 1% decrease in total ozone column. It depends on the wavelengths involved in the induction of skin cancer by UV radiation because, when ozone levels decrease, the shorter wavelengths in the UV-B range increase more steeply than the longer wavelengths. The optical amplification also depends on the pathlength of the UV radiation through the ozone layer. This makes the optical amplification to some extent dependent on geographical latitude, i.e. it increases from the equator to the poles.

The biological amplification factor indicates the percentage increase in skin cancer incidence caused by a 1% increase in the carcinogenically effective radiation. It is not directly proportional to the UV-B radiation, but follows a higher power of the radiation (Fears et al., 1977).

On the basis of a recently determined action spectrum for UV carcinogenesis in hairless mice (Slaper, 1987), it can be calculated using both amplification factors that with a 1% decrease in total ozone column, effective UV-B radiation will increase by 1.6%, the incidence of basal cell carcinomas by 2.7% and of squamous cell carcinomas by 4.6% (van der Leun, 1989). In many registries, basal cell carcinomas and squamous cell carcinomas are still considered together, as non-melanoma skin cancers. For a 1% depletion of ozone, the overall incidence of non-melanoma skin cancer would increase on average by about 3%. With a 5% decrease in stratospheric ozone, the incidence of basal cell carcinoma would increase by 14%, squamous cell carcinoma by 25%, and non-melanoma skin cancer in general by 16%. The lighter-skinned white populations of the world would be most affected, with 45 000 new cases of non-melanoma skin cancer per year (worldwide) following 1% stratospheric ozone depletion and 240 000 new cases annually following 5% ozone depletion (van der Leun, 1989).

The full importance of such numbers is difficult to define at present. Non-melanoma skin cancers, if detected early, have a very high cure rate. The current death rate is about 1% of known cases and reduced exposure (e.g., decreased everyday outdoor activities, decreased recreational sun exposure, use of more extensive protective clothing by farmers or other outdoor workers) could help to reduce skin cancer increases due to ozone depletion.

Sufficient evidence exists to show that sunlight also plays a role in the much more dangerous (often fatal) melanoma forms of skin cancer. A key uncertainty is the extent to which UV-B radiation, versus other sunlight components, may specifically contribute to the induction of cutaneous melanomas. The lack, until very recently, of any viable experimental animal model in which to study these light-activated skin pigment cell cancers has impeded progress. However, UV-B radiation has been shown to increase melanomas in two animal models (Setlow, 1988; Ley, 1988), and Kripke (1989) has noted that much greater growth of melanomas occurs when transplanted into the skin of UV-irradiated animals, suggesting that UV radiation may not only initiate melanomas but have promoter effects as well. Thus, it is prudent to consider melanoma skin cancers among potential health effects of ozone layer depletion (US EPA, 1987d).

9.4.2 Immunotoxic effects

Another potential effect of stratospheric ozone depletion is suppression of immune function. UV-B radiation appears to modify immune function in irradiated mice in different ways both locally at the point of skin irradiation and systemically. Locally-induced effects include impairment of Langerhans cells and abnormal endogenous cancer cells (Kripke, 1989; Stingl et al., 1983). After exposure to UV-B radiation, Langerhans cells no longer present antigens to the helper T-lymphocytes. Consequently when contact allergens are applied to UV-B-exposed skin, no contact allergy ensues. Instead, suppressor lymphocytes are activated, which prevent any subsequent immune response to the same antigen. Suppressor T-cell lymphocytes are normally involved in regulating the magnitude and duration of immune responses. Their activation by UV-B radiation prevents the development of natural immune responses against UV-B-induced skin cancers and thereby contributes to their growth and spread to other parts of the body (Daynes et al., 1986). In addition, circulation of UV-B-activated suppressor lymphocytes throughout the body, and an associated reduction in helper lymphocytes, results in a general, systemic suppression of certain immune functions. Thus, UV-B-irradiated mice not only fail to exhibit contact allergy responses to chemicals applied to irradiated skin but they also have impaired ability to respond to chemicals applied to non-irradiated skin. In addition, they have decreased lymphocyte-mediated immune responses to foreign substances injected under the skin (i.e. delayed hypersensitivity reactions) (Noonan et al., 1981). The systemic suppression of immune function due to UV radiation has been demonstrated to (a) occur in several animal species, (b) increase as a function of increasing UV-B dosage, and (c) persist beyond the initial period of UV exposure (Giannini, 1986; Howie et al., 1986; Kripke, 1988). Two studies of infectious agents showed that UV-B radiation resulted in suppressed immune response in mice to *Herpes simplex* that lasted for several months (Howie et al., 1986) and that UV-B-irradiated mice infected with *Leishmania* failed to exhibit the delayed hypersensitivity immune responses that were induced in non-irradiated mice (Giannini, 1986). However, insufficient evidence exists to make a quantitative estimation of the dependence of potential increases in the incidences of specific types of infectious diseases in humans on stratospheric ozone depletion.

9.4.3 Ocular effects

Adverse occular effects resulting from the exposure to UV-B radiation may also be the consequence of stratospheric ozone depletion (US EPA, 1987a; van der Leun, 1988,1989). One effect, "snow-blindness", is typically transient, lasting only a few days, but its possible increase at higher (snowier) latitudes is of interest. Of much more concern, however, is evidence that UV-B radiation increases cataract formation (Pitts et al., 1986) and the suggestion that 1% stratospheric ozone depletion would increase cataract prevalence by about 0.25 to 0.6% (roughly equivalent to about 24 000 to 57 000 more cases in the USA per year at present population levels) (US EPA, 1987a).

9.4.4 Effects on vitamin D synthesis

Skin exposure to UV-B radiation causes the formation of vitamin D_3. It has been suggested that some population segments that suffer from vitamin D deficiency, e.g., dark-skinned children in northern cities, children in families with strict macrobiotic diets, elderly people living mainly indoors (Aaron et al., 1974), and especially children, might gain some benefit from increased UV-B radiation resulting from stratospheric ozone depletion. Excess production of vitamin D_3 in other groups would be unexpected since its formation is self-limiting (Holick, 1981).

9.4.5 Exacerbation of photochemical smog formation and effects

Increased UV-B radiation would be expected to facilitate tropospheric ozone formation and acid aerosol formation, the latter due to increased surface hydrogen peroxide concentrations. The health effects associated with both tropospheric ozone and acid aerosols would then be expected to occur with greater frequency and severity (US EPA, 1986; Lippmann, 1989; Grant, 1989).

10. EVALUATION OF HUMAN HEALTH RISKS AND EFFECTS ON THE ENVIRONMENT

10.1 Evaluation of human health risks

10.1.1 Direct health effects resulting from exposure to fully halogenated chlorofluorocarbons

The kinetics and metabolism of chlorofluorocarbons are characterized by rapid pulmonary absorption and distribution. There is no indication of any accumulation. Metabolic transformation of the chlorofluorocarbons considered in this monograph is negligible, if it occurs at all. Therefore, toxic effects of metabolites are very unlikely. The acute toxicity of chlorofluorocarbons is very low, as demonstrated in studies on various animal species and by different routes of administration. It is characterized by effects on the heart, the respiratory system, and occasionally the liver. The effects are in accordance with the symptomatology observed in acute intoxications in humans.

After repeated exposure, comparable clinical symptoms can be observed. Alterations in the liver and kidney occur occasionally. In humans, CNS, cardiovascular and respiratory symptoms occur in cases of severe abuse and in uncontrolled or accidental occupational exposure. Under conditions of use involving short-term exposures of up to 1000 ppm, no adverse health effects would be expected.

An evaluation of the animal studies indicates no carcinogenic risk to human beings. This is underlined by the fact that the chlorofluorocarbons discussed in this monograph are devoid of genotoxicity in different mutagenic end-points and cell transformation. In a limited cohort study with 539 exposed workers, neither increased deaths nor increased tumour ratios were reported. Studies on the influence of reproduction (fertility, embryotoxicity, fetotoxicity, teratology) and general effects on developmental processes in experimental animals were consistently negative. Effects on human reproduction, including intrauterine and post-natal development, have not been reported.

Mean concentrations in the ambient air in urban/suburban areas of 3.4 µg/m³ for CFC-11 and 6 µg/m³ for CFC-12 have been measured. In remote/rural areas, the corresponding levels were 1.0 µg/m³ for CFC-11 and 1.6 µg/m³ for CFC-12.

These exposure levels are considered negligible in comparison with the concentrations of 25 000 to 50 000 µg/m³ (≈5000 to 10 000 ppm) that cause initial signs either of functional or morphological changes in laboratory animals. They are particularly negligible in comparison with the very high exposure levels that cause functional changes in humans.

10.1.2 Health effects expected from reduction of stratospheric ozone by chlorofluorocarbons

During the last decade, increasing concern has been focused on the consequences of a reduction of ozone in the upper atmosphere, with the concomitant increase of UV-B radiation at the surface of the earth. Model calculations predict, for the next 50 years, ozone depletions of between 1 and 10%, depending on the scenario used for the release of chlorofluorocarbons and other trace gases.

Among the effects on human health, the induction of non-melanoma skin cancer has been investigated extensively, both in human epidemiology and in animal experimental work. It is a generally accepted conclusion that the incidence of non-melanoma skin cancer will increase as a result of ozone depletion. An estimation based on recent data predicts that a decrease of atmospheric ozone by 1% would lead to an increase in the incidence of non-melanoma skin cancers by 3%. An ozone depletion by 5% would lead to an increase of the incidence by 16%. This latter increase would mean a worldwide increase of about 240 000 additional new patients with non-melanoma skin cancer per year, predominantly light skinned people.

Indications are increasing that UV-B radiation also plays a role in the induction and growth of cutaneous melanomas, a more dangerous type of skin cancer. Uncertainty, however, especially with regard to the dose-effect relationship, makes quantitative predictions very difficult. The possibility of an increase in cutaneous melanoma should therefore be taken into consideration.

The immune system of experimental animals is suppressed in specific ways by UV-B radiation. This results in a decreased resistance to implanted UV-B-induced tumours and an increased growth of such tumours in mice, in the suppression of sensitization by contact allergens, and the response to allergens in sensitized animals. It also results in the impairment of the immune response against certain infectious agents; this has been demonstrated for *Herpes simplex* and *Leishmania* sp. There are indications that similar suppression of the immune response by UV-B radiation may occur in humans. The antigen-presenting Langerhans cells in the skin are damaged and allergic responses are depressed. Although much still has to be learned through further research, the possibility that immune suppression effects and a consequent increase in the incidence of some infectious diseases might occur as a result of stratospheric ozone depletion should not be ignored.

There are indications that UV-B radiation increases cataract formation, an important cause of blindness especially in areas with limited medical facilities.

10.2 Effects on the environment

Other than the theory that chlorofluorocarbons contribute to the "greenhouse" effect, there is no evidence available of other direct ecological effects produced by the chlorofluorocarbons discussed in this monograph.

Studies addressing the effects of UV-B radiation on plants have concentrated on crop plants and have usually been conducted at temperate latitudes. This represents only a small portion of the major ecosystems of the world. Although there are many uncertainties resulting from the complexities of the experiments, the data now available suggest that crop yields are potentially vulnerable to increased levels of solar UV-B radiation. Out of more than 200 species and cultivars screened for UV tolerance, about two-thirds have been found to be sensitive. The most sensitive plant groups include crops related to peas and beans, melons, mustard, and cabbage. Members of the grass family are generally less sensitive.

Experimental evidence indicates that there is some degree of tolerance to UV-B radiation in the gene pool. This is based on the high degree of variation in sensitivity to UV radiation among crop cultivars. The genetic basis for the sensitivity has yet to be determined.

The effect of enhanced levels of UV-B radiation on the quality of crops has been studied. The protein and oil content of selected cultivars of soybean seeds were reduced by up to 10% when plants were exposed to UV levels simulating a 25% ozone depletion.

Studies regarding the effects of UV-B radiation on forest productivity are limited. There are results only for seedlings, and they are for levels of exposure equivalent to a 40% ozone reduction. These studies showed a reduction in growth and photosynthesis following the exposure of loblolly pine seedlings. There is experimental evidence that increased UV-B radiation levels can cause shifts in community structure.

Exposure to UV-B radiation has been shown to affect both plant and animal components of marine ecosystems. These effects include decreases in fecundity, growth, survival, and other parameters.

10.3 Conclusions

The available toxicological data on the fully halogenated chlorofluorocarbons reviewed in this monograph show a low acute and chronic toxicity and indicate no mutagenic or carcinogenic potential. The human health risks are mainly confined to occasional high exposures that may occur when handling these substances. In contrast, the indirect effects on human beings from the accumulation of these substances in the stratosphere may lead to substantial effects on human health, mainly due to the depletion of stratospheric ozone resulting in an increase in effects from UV-B radiation. The projected increase in the incidence of non-melanoma skin cancers, a possible increase in melanoma skin cancers, and immunotoxic and ocular effects all lead to the conclusion that immediate and effective international cooperation is necessary to reduce further stratospheric ozone depletion.

11. RECOMMENDATIONS

1. The toxicity data base for some chlorofluorocarbons, especially those containing hydrogen, is inadequate for quantitative risk assessment. Additional information on the chronic toxicity, carcinogenicity, and teratogenicity/reproductive effects of these compounds, especially by inhalation exposure, is needed.

2. An assessment of the effects of increased UV-B radiation is summarized in Table 16.

Table 16. Potential effects of increased UV-B radiation resulting from decreased stratospheric ozone[a]

Effects	State of knowledge	Potential global impact
Skin cancer	Moderate to high	Moderate
Immune system	Low	High
Cataracts	Moderate	Low[b]
Plant life[c]	Low	High
Aquatic life[c]	Low	High
Climate impacts[d]	Moderate	Moderate
Ambient ozone	Moderate	Low[e]

[a] Modified from SAB-EC-87-025 *Review of EPA's Assessment of the Risks of Stratospheric Modification* by the Stratospheric Ozone Subcommittee, Science Advisory Board, US Environmental Protection Agency, March, 1987.
[b] A more recent consideration of the influence of ozone depletion on the incidence of cataracts suggests that the impact in this respect may be more serious (US EPA, Assessing the Risks of Trace Gases that can modify the Stratosphere, Chapter 10, December 1987).
[c] See section 6.
[d] Contribution of both stratospheric ozone depletion itself and gases causing such depletion to climatic changes, including sea level rise.
[e] Impact could be high in selected urban or rural areas typified by local or regional scale surface-level ozone air pollution problems.

More research is needed in those areas where knowledge is lacking and the potential global impact is high. These include the following eight specific areas of future research and assessment that are especially important for understanding and dealing with stratospheric ozone depletion effects on human health:

- the investigation of mechanisms of immunosuppression in animal models and humans;
- the identification of infectious diseases that include a stage or process that could be worsened by exposure to UV-B radiation, and the development of models to explain these diseases;
- the investigation of wavelength dependence and the development of dose-response information for humans concerning the effects of UV-B exposure on the incidence of infectious diseases;
- the determination of the impact of UV-B immunosuppression on vaccination efficacy;
- the clarification of the role of immunological changes in the induction of melanomas and non-melanoma skin cancers by UV radiation;
- the determination of the action spectra and dose-effect relationships for the induction of the various types of melanoma by UV radiation;
- the establishment of a better definition of the action spectra for the induction of squamous cell carcinoma, and especially basal cell carcinoma, by UV radiation;
- the investigation of the biology and epidemiology of cataracts, and of methods to reduce the risk of eye diseases.

3. The use of CFC-11 and CFC-12 as propellants for the disinsection of aircraft by aerosol sprays is still recommended by some authorities. There is urgent need for a new non-ozone-depleting, non-flammable, safe, non-irritant propellant for this use, since the older propellants are already banned in many countries.

4. Effective international cooperation is necessary to reduce future stratospheric ozone depletion, and for this purpose cuts in the emission of ozone-depleting chlorofluorocarbons of at least 80-90% are necessary. The first priority is to find substitutes and the second to devise adequate disposal procedures for existing waste chlorofluorocarbons. It is recommended that all countries take steps to reduce the use of chlorofluorocarbons with high stratospheric ozone-depletion potential.

REFERENCES

ADIR, J., BLAKE, D.A., & MERGNER, G.M. (1975) Pharmacokinetics of fluorocarbon 11 and 12 in dogs and humans. *J. clin. Pharmacol.*, 15: 760-770.

ALLEN & HANBURYS, LTD (1971) *An investigation of possible cardiotoxic effects of the aerosol propellants, arctons 11 and 12*, Vol. 1 (Unpublished report, courtesy of D. Jack, Managing Director, Allen & Hanburys, Ltd).

AMIN, Y.M., THOMPSON, E.B., & CHIOU, W.L. (1979) Fluorocarbon aerosol propellants. XIII. Correlation of blood levels of trichloromonofluoromethane to cardiovascular and respiratory responses in anesthetized dogs. *J. pharm. Sci.*, 68(2): 160-163.

ANGERER, J., SCHROEDER, B., & HEINRICH, R. (1985) Exposure to fluorotrichlormethane (R-11). *Int. Arch. occup. environ. Health*, 56(1): 67-72.

AVIADO, D.M. (1978) Effects of fluorocarbons, chlorinated solvents, and inosine on the cardiopulmonary system. *Environ. Health Perspect.*, 26: 207-215.

AVIADO, D.M. & BELEJ, M.A. (1974) Toxicity of aerosol propellants on the respiratory and circulatory systems. I. Cardiac arrhythmia in the mouse. *Toxicology*, 2: 31-42.

AVIADO, D.M. & MICOZZI, M.S. (1981) Fluorine-containing organic compounds in industrial hygiene and toxicology. In: Clayton, G.D. & Clayton, F.E., ed. *Patty's industrial hygiene and toxicology*, 3rd revised ed., New York, John Wiley and Sons, Vol. 2B, pp. 3071-3115.

AVIADO, D.M. & SMITH, D.G. (1975) Toxicity of aerosol propellants in the respiratory and circulatory systems. VIII. Respiration and circulation in primates. *Toxicology*, 3: 241-252.

AZAR, A., ZAPP, J.A., REINHARDT, C.F., & STOPPS, G.J. (1971) Cardiac toxicity of aerosol propellants. *J. Am. Med. Assoc.*, 215(9): 1501-1502.

AZAR, A., REINHARDT, C.F., MAXFIELD, M.E., SMITH, P.E., Jr, & MULLIN, L.S. (1972) Experimental human exposure to fluorocarbon (sic) 12 (dichlorodifluoromethane). *Am. Ind. Hyg. Assoc. J.*, 33(4): 207-216.

AZAR, A., TROCHIMOWICZ, H.J., TERRILL, J.B., & MULLIN, L.S. (1973) Blood levels of fluorocarbon related to cardiac sensitization. *Am. Ind. Hyg. Assoc. J.*, 34: 102-109.

BARTSCH, H., MALAVEILLE, C., CAMUS, A.-M., MARTEL-PLANCHE, G., BRUN, G., HAUTEFEUILLE, A., SABADIE, N., BARBIN, A., KUROKI, T., DREVON, C., PICCOLI, C., & MONTESANO, R. (1980) Validation and comparative studies on 180 chemicals with *S. typhimurium* strains and V79 Chinese hamster cells in the presence of various metabolizing system. *Mutat. Res.*, 76: 1-50.

BASS, M. (1970) Sudden sniffing death. *J. Am. Med. Assoc.*, 212(12): 2075-2079.

BEGGS, C.J., SCHNEIDER-ZIEBERT, V., & WELLMANN, E. (1986) UV-B radiation and adaptive mechanisms in plants. In: Worrest, R.C. & Coldwell, M.M., ed. *Stratospheric ozone reduction, solar UV radiation and plant life*, Heidelberg, Springer-Verlag, pp. 235-250.

BELEJ, M.A., SMITH, D.G., & AVIADO, D.M. (1974) Toxicology of aerosol propellants in the respiratory and circulatory systems. IV. Cardiotoxicity in the monkey. *Toxicology*, 2: 381-395.

BEYSCHLAG, W., BARNES, P.W., FLINT, S.D., & CALDWELL, M.M. (1988) Enhanced UV-B irradiation has no effect on photosynthesis characteristics of wheat *(Triticum aestivum L.)* and wild oat *(Avena falva L.)* under greenhouse and field conditions. *Photosynthetica*, 22: 31-37.

BLAKE, D.A. & MERGNER, G.W. (1974) Inhalation studies on the biotransformation and elimination of ^{14}C-trichlorofluoromethane and ^{14}C-dichlorodifluoromethane in beagles. *Toxicol. appl. Pharmacol.*, 30: 396-407.

BOWER, F.A. (1973) *Nomenclature and chemistry of fluorocarbon compounds*, Springfield, Virginia, US National Technical Information Service, 9 pp (NTIS AD Report No. 751423).

BRICE, K.A., DERWENT, R.G., EGGLETON, A.E.J., & PENKETT, S.A. (1982) Measurements of trichlorofluoromethane and tetrachloromethane at Harwell over the period January 1975 - June 1981 and the atmospheric lifetime of trichlorofluoromethane. *Atmos. Environ.*, 16(11): 2543-2554.

BRODZINSKY, R. & SINGH, H.B. (1982) *Volatile organic chemicals in the atmosphere: an assessment of available data*, Research Triangle Park, North Carolina, US Environmental Protection Agency, Office of Research and Development, Environmental Sciences Research Laboratory (Final report on Contract No. 68-022-3452).

BRUGNONE, F., PERBELLINI, L., & APOSTOLI, P. (1984) Blood concentration of solvents in industrial exposure. *Collect. Méd. lég. Toxicol. méd.*, 125: 165-168.

BRUNER, F., CRESCENTINI, G., & MANGANI, F. (1981) Determination of halocarbons in air by gas chromatograph-high-resolution mass spectrometry. *Anal. Chem.*, 53: 798-801.

BULLISTER, J.L. & WEISS, R.F. (1983) Anthropogenic chlorofluoromethanes in the Greenland and Norwegian seas. *Science*, 221(4607): 265-268.

CALLAHAN, M.A., SLIMAK, M.W., GABEL, N.W., MAY, I., FOWLER, C., FREED, R., JENNINGS, P., DURFEE, R., WHITMORE, F., MAESTRI, B., MABEY, W., HOLT, B., & GOULD, C. (1979) *Water-related environmental fate of 129 priority pollutant. II. Halogenated aliphatic compounds, halogenated ethers, monocyclic aromatics, phthalate esters, polycyclic aromatic hydrocarbons, nitrosamines, and miscellaneous compounds*, Washington, DC, US Environmental Protection Agency (EPA-440/4-79-029b).

CALLIGHAN, J.A. (1971) Thermal stability data on six fluorocarbons. *Heat. Piping air Cond.*, **43**: 119-126.

CAMPBELL, D.D., LOCKEY, J.E., PETAJAN, J., GUNTER, B.J., & ROM, W.N. (1986) Health effects among refrigerator repair workers exposed to fluorocarbons. *Br. J. ind. Med.*, **43**: 107-111.

CARTER, V.L. (1970) *Data given in discussion of Carter et al. (1970)*, Springfield, Virginia, National Technical Information Service, p. 315 (NTIS AD Report No. 727523).

CARTER, V.L., CHIKOS, P.M., MACEWEN, J.D., & BACK, K.C. (1970) *Effects of inhalation of Freon 113 on laboratory animals*, Springfield, Virginia, National Technical Information Service, 18 pp (NTIS AD Report No. 727523).

CAUJOLLE, F. (1964) Comparative toxicity of refrigerants. *Bull. Inst. Intern. Froid*, **1**: 21-54.

CHIOU, W.L. & NIAZI, S. (1973) A simple and ultra-sensitive head-space gas chromatographic method for the assay of fluorocarbon propellants in blood. *Res. Commun. chem. Pathol. Pharmacol.*, **6**(2): 481-498.

CLARK, D.G. & TINSTON, D.J. (1972a) The influence of fluorocarbon propellants on the arrhythmogenic activities of adrenaline and isoprenaline. *Proc. Eur. Soc. Study Drug Toxicity*, **13**: 212-217.

CLARK, D.G. & TINSTON, D.J. (1972b) Cardiac effects of isoproterenol, hypoxia, hypercapnia, and fluorocarbon propellants and their use in asthma inhalers. *Ann. Allergy*, **30**(9): 536-541.

CLAYTON, J.W., Jr (1962) The toxicity of fluorocarbons with special reference to chemical constitution. *J. occup. Med.*, **4**(5): 262-273.

CLAYTON, J.W., Jr (1966) The mammalian toxicology of organic compounds containing fluorine. *Handb. exp. Pharmakol.*, **20**: 459-500.

CLAYTON, J.W., Jr (1967a) Fluorocarbon toxicity and biological action. *Fluorine chem. Rev.*, **1**(2): 197-252.

CLAYTON, J.W., Jr (1967b) Fluorocarbon toxicity: past, present, future. *J. Soc. Cosmet. Chem.*, **18**: 333-350.

CLAYTON, J.W., Jr (1970) Highlights of fluorocarbon toxicology. In: Sunderman, F.W., ed. *Laboratory diagnosis of diseases caused by toxic agents*, St. Louis, Missouri, Warren H. Green, Inc., Chapter 21A, pp. 199-214.

CLAYTON, J.W., Jr, SHERMAN, H., MORRISON, S.D., BARNES, J.R., & HOOD, D.B. (1964) Toxicity studies on 1,2-difluorotetrachloroethane (Freon-112) and 1,1-difluorotetrachloroethane (Freon-112A). *Toxicol. appl. Pharmacol.*, **6**: 342.

CLAYTON, J.W., Jr, HOOD, D.B., NICK, M.S., & WARITZ, R.S. (1966) Inhalation studies on chloropentafluoroethane. *Am. Ind. Hyg. Assoc. J.*, 27: 234-238.

CMA (1986) *CMA news release: production, sales, and calculated release of CFC-11 and CFC-12 through 1985*, Washington, DC, Chemical Manufacturers Association, Fluorocarbon Program Panel.

CMR (1975) *Chemical profile: fluorocarbons*, New York, Chemical Marketing Reporter.

CMR (1978) *Chemical profile: fluorocarbons*, New York, Chemical Marketing Reporter.

CMR (1981) *Chemical profile: fluorocarbons*, New York, Chemical Marketing Reporter.

CMR (1986) *Chemical profile: fluorocarbons*, New York, Chemical Marketing Reporter.

COATE, W.B., RAPP, R.W., ANDERSON, J., & CHARM, J. (1979) Inhalation toxicity of monochloromonofluoromethane. *Toxicol. appl. Pharmacol.*, 48(A109): 79.

COLLINS, G.F., BARTLETT, F.E., TURK, A., EDMONDS, S.M., & MARK, H.L. (1965) A preliminary evaluation of gas air tracers. *J. Air Pollut. Control Assoc.*, 15: 109.

COLLINS, G.G. & UTLEY, D. (1972) Simple membrane inlet for direct sampling of organic pollutants in the atmosphere by mass spectrometry. *Chem. Ind.*, 2: 84.

COX, P.J., KING, L.J., & PARKE, D.V. (1972a) A study of the possible metabolism of trichlorofluoromethane. *Biochem. J.*, 130(1): 13P-14P.

COX, P.J., KING, L.J., & PARKE, D.V. (1972b) A comparison of the interactions of trichlorofluoromethane and carbon tetrachloride with hepatic cytochrome P-450. *Biochem. J.*, 130: 87P.

COX, P.J., DERWENT, R.G., EGGLETON, A.E.J., & LOVELOCK, J.E. (1976) Photochemical oxidation of halocarbons in the troposphere. *Atmos. Environ.*, 10: 305-308.

CRESCENTINI, G. & BRUNER, F. (1979) Evidence for the presence of Freon 21 in the atmosphere. *Nature (Lond.)*, 279: 311-312.

CRESCENTINI, G., MANGANI, F., MASTROGIACOMO, A.R., CAPPIELLO, A., & BRUNER, F. (1983) Fast determination of some halocarbons in the atmosphere by gas chromatography-high-resolution mass spectrometry. *J. Chromatogr.*, 280(1): 146-151.

CULIK, R. & SHERMAN, H. (1973) *Teratogenic study in rats with dichlorodifluoromethane (Freon 12)*, Newark, Delaware, Haskell Laboratories, 10 pp (Medical Research Project No. 1388; Report No. 206-73) (Unpublished, courtesy of Du Pont de Nemours & Co.).

CUNNOLD, D.M., PRINN, R.G., RASMUSSEN, R.A., SIMMONDS, P., ALYEA, F., CARDELINO, C., CRAWFORD, A., FRASER, P., & ROSEN, R. (1983a) The atmospheric lifetime experiment. III. Lifetime methodology and applications to 3 years of trichlorofluoromethane data. *J. geophys. Res.*, 88(13): 8379-8400.

CUNNOLD, D.M., PRINN, R.B., SIMMONDS, A., RASMUSSEN, R., ALYEA, F., CARDELINO, C., & CRAWFORD, A. (1983b) The atmospheric lifetime experiment. IV. Results for dichlorodifluoromethane based on three years data. *J. geophys. Res.*, 88(13): 8401-8414.

DAMKAER, D.M. (1982) Possible influence of solar UV radiation in the evolution of marine zooplankton. In: Calkins, J., ed. *The role of solar ultraviolet radiation in marine ecosystems*, New York, London, Plenum Press, pp. 701-706.

DAMKAER, D.M., DEY, D.M., HERON, G.A., & PRENTICE, E.F. (1980) Effects of UV-B radiation on near-surface zooplankton of Puget Sound. *Oecologia*, 44: 149-158.

DANIELS, S., PATON, W.D.M., & SMITH, E.B. (1979) The effects of some hydrophobic gases on the pulmonary surfactant system. *Br. J. Pharmacol.*, 65(2): 229-235.

DESOILLE, H., TRUFFERT, L., BOURGUIGNON, A., DELAVIERRE, P., PHILBERT, M., & GIRARD-WALLON, C. (1968) Experimental study on the toxicity of trichlorotrifluoroethane (Freon 113). I. *Arch. Mal. prof. Méd. Trav. Sécur. soc.*, 29(7-8): 381-388.

DICKSON, A.G. & RILEY, J.P. (1976) The distribution of short-chain halogenated aliphatic hydrocarbons in some marine organisms. *Mar. Pollut. Bull.*, 7(9): 167-169.

DOHERTY, R.E. & AVIADO, D.M. (1975) Toxicity of aerosol propellants in the respiratory and circulatory systems. VI. Influence of cardiac and pulmonary vascular lesions in the rat. *Toxicology*, 3: 213-224.

DOLLERY, C.T., DRAFFAN, G.H., DAVIES, D.S., WILLIAMS, F.M., & CONOLLY, M.E. (1970) Blood concentrations in man of fluorinated hydrocarbons after inhalation of pressurized aerosols. *Lancet*, 2(7684): 1164-1166.

DOUCET, J., SAUVAGEAU, P., & SANDORFY, C. (1973) Vacuum ultraviolet and photoelectron spectra of fluoro-chloro derivatives of methane. *J. chem. Phys.*, 58: 3708-3716.

DOUCET, J., SAUVAGEAU, P., & SANDORFY, C. (1974) The photoelectron and far-ultraviolet absorption spectra of chlorofluoro derivatives of ethane. *J. chem. Phys.*, 62(2): 355-359.

DOWNING, R.C. (1966) Fluorinated hydrocarbons. In: Standen, A., ed. *Kirk-Othmer encyclopedia of chemical technology*, 2nd ed., New York, John Wiley and Sons, Vol. 9, pp. 739-751.

DU PONT (1965) *Toxicity studies with 1,1,2-trichloro-1,2,2-trifluoroethane*, Wilmington, Delaware, E.I. Du Pont de Nemours & Co., pp. 1-12 (Freon Technical Bulletin No. S-24).

DU PONT (1980a) *Toxicity studies with 1,1,2-trichloro-1,2,2-trifluoroethane*, Wilmington, Delaware, E.I. Du Pont Nemours & Co (Technical Bulletin No. S-24).

DU PONT (1980b) *Freon fluorocarbons properties and applications. B-2*, Wilmington, Delaware, E.I. Du Pont de Nemours & Co.

EDDY, C.W. & GRIFFITH, F.D. (1971) *Metabolism of dichlorodifluoromethane-C^{14} by rats*. Presented at the American Industrial Hygiene Association Conference, Toronto, Canada, May 1971, New York, American National Standards Institute and Akron, Ohio, American Industrial Hygiene Association.

EDLING, C. & OLSON, C.G. (1988) [*Health risks with exposure to freons*], Uppsala, Sweden, Department of Occupational Medicine, University Hospital (in Swedish).

EGLE, J.L., BORZELLECA, J.F., & PUTNEY, J.W., Jr (1972) Cardiac function in mice following exposure to haloalkane propellants alone and in combination with bronchodilators. In: *Proceedings of the 3rd Annual Conference on Environmental Toxicology, Fairborn, Ohio, 25-27 October*, Springfield, Virginia, National Technical Information Service, pp. 239-247 (NTIS AD Report No. 773766; AMRL-TR-72-130; Paper No. 15).

EMBER, L.R., LAYMAN, P.L., LEPKOWSKI, W., & ZURER, P.S. (1986) The changing atmosphere. *Chem. eng. News*, 24 November: 14-64.

EPSTEIN, S.S., JOSHI, S., ANDREA, J., CLAPP, F., FALK, H., & MANTEL, N. (1967a) Synergistic toxicity and carcinogenicity of "freons" and piperonyl butoxide. *Nature (Lond.)*, 214(5087): 526-528.

EPSTEIN, S.S., ANDREA, J., CLAPP, P., MACKINTOSH, D., & MANTEL, N. (1967b) Enhancement by piperonyl butoxide of acute toxicity due to freons, benzo(α)pyrene, and griseofulvin in infant mice. *Toxicol. appl. Pharmacol.*, 11: 442-448.

EPSTEIN, S.S., ARNOLD, E., ANDREA, J., BASS, W., & BISHOP, Y. (1972) Detection of chemical mutagens by the dominant lethal assay in the mouse. *Toxicol. appl. Pharmacol.*, 23: 288-325.

FAGAN, D.G., FORREST, J.B., ENHORNING, G., LAMPREY, M., & GUY, J. (1977) Acute pulmonary toxicity of a commercial fluorocarbon-lipid aerosol. *Histopathology*, 1(3): 209-223.

References

FLOWERS, N.C. & HORAN, L.H. (1972) Effects of respiratory acidosis on the cardiac response to aerosol inhalation. *Clin. Res.*, 20: 619.

FLOWERS, N.C., HAND, R.C., & HORAN, L.G. (1975) Concentrations of fluoroalcanes associated with cardiac conduction system toxicity. *Arch. environ. Health*, 30(7): 353-360.

FOLTZ, V.C. & FUERST, R. (1974) Mutation studies with *Drosophila melanogaster* exposed to four fluorinated hydrocarbon gases. *Environ. Res.*, 7: 275-285.

FRASER, P.J., HYSON, P., ENTING, I.G., & PEARMAN, G.I. (1983) Global distribution and southern hemispheric trends of atmospheric trichlorfluoromethane. *Nature (Lond.)*, 302(5910): 692-695.

FRIEDMAN, S.A., CAMMARATO, M., & AVIADO, D.M. (1973) Toxicity of aerosol propellants on the respiratory systems. II. Respiratory and bronchopulmonary effects in the rat. *Toxicology*, 1: 345-355.

FULLER, B., HUSKON, J., KORNREICH, M., OUELLETTE, R., THOMAS, L., & WALKER, P. (1976) *Preliminary scoring of selected organic air pollutants*, Research Triangle Park, North Carolina, US Environmental Protection Agency, Office of Air and Waste Management, pp. 4-39 (EPA-450/3-77-008a).

FURUYA, M. (1979) [Experimental chlorofluorohydrocarbon poisoning.] *Tokyo Jikeikai Ika Daigaku Zasshi*, 94(6): 1201-1214 (in Japanese).

GARB, S. & CHENOWETH, M.B. (1948) Studies on hydrocarbon-epinephrine induced ventricular fibrillation. *J. Pharmacol. exp. Ther.*, 94: 12-18.

GARRETT, S. & FUERST, R. (1974) Sex-linked mutations in *Drosophila* after exposure to various mixtures of gas atmospheres. *Environ. Res.*, 7: 286-293.

GELBICOVA-RUZICKOVA, J., NOVAK, J., & JANAK, J. (1972) Application of the method chromatographic equilibrium to air pollution studies. The determination of minute amounts of halothane in the atmosphere of an operating theatre. *J. Chromatogr.*, 64: 15-23.

GELLER, I., HARTMAN, R.J., Jr, & MENDEZ, V.M. (1977) *Evaluation of performance impairment by spacecraft contaminants*, San Antonio, Texas, Southwest Foundation for Research and Education, 71 pp (Contract Report No. NASA-CR-151478).

GOLD, N.G. & CALDWELL, M.M. (1983) The effects of ultraviolet-B radiation on plant competition in terrestrial ecosystems. *Physiol. Plant.*, 58: 435-444.

GRAFF-LONNEVIG, V. (1979) Diurnal expiratory flow after inhalation of Freons and Genoterol in childhood asthma. *J. Allergy clin. Immunol.*, 64: 534-538.

GRANT, L.D. (1989) Health effects issues associated with regional and global air pollution problems. In: *Proceedings of the World Conference on the Changing Atmosphere, Toronto, June*, Geneva, World Meteorology Organization and Ottawa, Canada Department of the Environment, pp. 243-270.

GREEN, T. (1983) The metabolic activation of dichloromethane and chlorofluoromethane in a bacterial mutation assay using *Salmonella typhimurium*. *Mutat. Res.*, 118(4): 277-288.

GREENBERG, L.A. & LESTER, D. (1950) Toxicity of the tetrachlorodifluoroethanes. *Arch. ind. Hyg. occup. Med.*, 2: 345-347.

GRIFFIN, T.B., BYARD, J.L., & COULSTON, F. (1972) Toxicological responses to halogenated hydrocarbons. In: *Appraisal of halogenated fire extinguishing agents. Proceedings of a Symposium, Washington, DC, 11-12 April*, Washington, DC, US National Academy of Science, pp. 136-147.

GUICHERIT, R. & SCHULTING, F.L. (1985) The occurrence of organic chemicals in the atmosphere of the Netherlands. *Sci. total Environ.*, 43(3): 193-219.

HAJ, M., BURSTEIN, Z., HORN, E., & STAMLER, B. (1980) Perforation of the stomach due to trichlorofluoromethane (Freon 11) ingestion. *Isr. J. med. Sci.*, 16(5): 392-394.

HALL, D.W. (1984) *Volatile organic contamination in an alluvial aquifer, Southington, Connecticut.* Presented at the National Conference and Exhibit on Hazardous Wastes in Environmental Emergencies: Management, Prevention, Clean-up, and Control, pp. 190-197.

HAMILTON, J.M. (1962) The organic fluorochemical industry. *Adv. fluorine Chem.*, 3: 117-180.

HAMMITT, J.K. (1986) *Product uses and market trends for potential ozone-depleting substances, 1985-2000*, Santa Monica, California, Rand Corporation (Report No. R-3386-EPA).

HANAI, Y., KATOU, T., & IIZUKA, T. (1984) [Automatic analysis of halogenated hydrocarbons in the air.] *Yokohama Kokuritsu Daigaku Kankyo Kagaku Kenkyu Senta Kiyo*, 11(1): 17-27 (in Japanese).

HANSCH, C., VITTORIA, A., SILIPO, C., & JOW, P.Y.C. (1975) Partition coefficients and the structure-activity relationship of the anesthetic gases. *J. med. Chem.*, 18(6): 546-548.

HANST, P.L. (1978) Noxious trace gases in the air. Part II. Halogenated pollutants. *Chemistry*, 51(2): 6-12.

HARRIS, W.S. (1972) Cardiac effects of halogenated hydrocarbons. In: *Appraisal of halogenated fire extinguishing agents. Proceedings of a Symposium, Washington, DC, 11-12 April*, Washington, DC, US National Academy of Science, pp. 114-126.

HARRIS, W.S. (1973) Aerosol propellants are toxic to the heart. *J. Am. Med. Assoc.*, 223: 1508-1509.

HARRIS, W.S., KILEN, S.M., TAYLOR, G.J., & LEVITSKY, S. (1971) Evidence from animals and man that freon depresses myocardial contractility. *Circulation*, 44(Suppl. II): 119.

References

HAWLEY, G.G., ed. (1981) *The condensed chemical dictionary,* 10th ed., New York, Van Nostrand Reinhold Company, pp. 236, 229, 244, 335-336, 229, 471, 1006, 1042.

HAYS, H.W. (1972) Etiology of cardiac arrhythmias. In: *Proceedings of the 3rd Annual Conference on Environmental Toxicology, Fairborn, Ohio, 25-27 October,* Springfield, Virginia, National Technical Information Service, pp. 173-183 (NTIS AD Report No. 7737166).

HAZELTON LABORATORIES (1967a) *Reproduction study in rabbits,* Vienna, Virginia, Hazelton Laboratories (MRO-1962-001, HLO-0258-67) (Unpublished data of E.I. Du Pont de Nemours & Co.).

HAZELTON LABORATORIES (1967b) *Teratology study in rabbits,* Vienna, Virginia, Hazelton Laboratories (MRO-1015, HLO-242. Final report) (Unpublished data of E.I. Du Pont de Nemours & Co.).

HERMANN, H. & VIAL, J. (1935) Nouvelles syncopes cardiaques par association toxique de l'adrénaline et de divers produits organiques volatiles. *C.R. Soc. Biol. (Paris),* **119**: 1316-1317.

HESTER, N.E., STEPHENS, E.R., & TAYLOR, O.C. (1974) Fluorocarbons in the Los Angeles Basin. *J. Air Pollut. Control Assoc.,* **24**(6): 591-595.

HOLICK, M.F. (1981) The cutaneous photosynthesis of pre-vitamin D3: a unique photoendocrine system. *J. invest. Dermatol.,* **77**(1): 51-58.

HORVATH, A.L. (1982) *Halogenated hydrocarbons: solubility-miscibility with water,* New York, Marcel Dekker, p. 493.

HUNTER, J.R., KAUPP, S.E., & TAYLOR, J.H. (1982) Assessment of the effects of UV radiation on marine fish larvae. In: Calkins, J., ed. *The role of solar ultraviolet radiation in marine ecosystems,* New York, London, Plenum Press, pp. 459-497.

HUNTER-SMITH, R.J., BALLS, P.W., & LISS, P.S (1983) Henry's Law constants and the air-sea exchange of various low-molecular-weight halocarbon gases. *Tellus,* **35B**(3): 170-176.

HUSKINS, C.W., TARRANT, P., BRUESH, J.F., & PADBURY, J.J. (1951) Thermal dehydrohalogenation of some chlorofluoroethanes. *Ind. eng. Chem.,* **43**: 1253-1256.

IMBUS, H.R. & ADKINS, C. (1972) Physical examinations of workers exposed to trichlorotrifluoroethane. *Arch. environ. Health,* **24**(4): 257-261.

ISAKSEN, I.S.A. & STORDAL, F. (1981) The influence of man on the ozone layer: readjusting the estimates. *Ambio,* **10**(1): 9-17.

JACK, D. (1971) Sniffing syndrome. *Br. med. J.,* **2**: 708-709.

JAYANTY, R.K.M., SIMONAITIS, R., & HEICKLEN, J. (1975) The photolysis of chlorofluoromethanes in the presence of O_2 or O_3 at 213.9 nm and their reactions with O(1D). *J. Photochem.,* **4**: 381-398.

JENKINS, L.J., Jr, JONES, R.A., COON, R.A., & SIEGEL, J. (1970) Repeated and continuous exposures of laboratory animals to trichlorofluoromethane. *Toxicol. appl. Pharmacol.*, **16**: 133-142.

JOHNSON, W.E., FRASER, J.H., GIBSON, W.E., MODICA, A.P., & GROSSMAN, G. (1972) *Spray freezing, decanting, and hydrolysis as related to secondary refrigerant freezing*, Springfield, Virginia, US Department of the Interior, Office of Saline Water National Technical Information Service, (Research and Development Program Report No. 72-786; NTIS PB Report No. 215 036).

KAISER, K.L.E., COMBA, M.E., & HUNEAULT, H. (1983) Volatile halocarbon contaminants in the Niagara River and in Lake Ontario. *J. Great Lakes Res.*, **9**(2): 212-223.

KEHOE, R.A. (1943) *Report on human exposure to dichlorodifluoromethane in air*, Cincinnati, Ohio, Kettering Laboratory (Unpublished report).

KELLY, D.P., CULIK, R., TROCHIMOWICZ, H.J., & FAYERWEATHER, W.E. (1978) Inhalation teratology studies on three fluorocarbons. *Toxicol. appl. Pharmacol.*, **45**: 293.

KELLY, J.R. (1986) How might enhanced levels of solar UV-B radiation affect marine ecosystems? In: Titus, J.G., ed. *Effects of changes in stratospheric ozone and global climate*, Washington, DC, US Environmental Protection Agency and Nairobi, United Nations Environment Programme, Vol. 2, pp. 237-251.

KHALIL, M.A.K. & RASMUSSEN, R.A. (1983) Gaseous tracers of arctic haze. *Environ. Sci. Technol.*, **17**: 157-164.

KILEN, S.M. & HARRIS, W.S. (1972) Direct depression of myocardial contractility by the aerosol propellant gas, dichlorodifluoromethane. *J. Pharmacol. exp. Ther.*, **183**(2): 245-255.

KILEN, S.M. & HARRIS, W.S. (1976) Effects of hypoxia and Freon 12 on mechanics of cardiac contractions. *Am. J. Physiol.*, **230**(6): 1701-1707.

KRAHN, D.F., BARSKY, F.C., & MCCOOEY, K.T. (1982) CHO/HGPRT mutation assay: evaluation of gases and volatile liquids. *Environ. Sci. Res.*, **25**: 91-103.

KRIPKE, M.L. (1989) Health effects of stratospheric ozone depletion: an overview. In: Schneider, T., Lee, S.D., Grant, L.D., & Walters, G., ed. *Atmospheric ozone research and its policy implications: Third US-Dutch International Symposium, Nijmegen, The Netherlands, May 1988*, Amsterdam, Oxford, New York, Elsevier Science Publishers, pp. 795-802.

LEE, I.P. & SUZUKI, K. (1981) Studies on the male reproductive toxicity of Freon 22. *Fundam. appl. Toxicol.*, **1**(3): 266-270.

LESSARD, Y. & PAULET, G. (1985) Mechanism of liposoluble drugs and general anesthetic's membrane action: action of difluoromethane (FC 12) on different types of cardiac fibers isolated from sheep hearts. *Cardiovasc. Res.*, **19**(8): 465-473.

LESSARD, Y. & PAULET, G. (1986) A proposed mechanism for cardiac sensitisation: Electrophysiological study of effects of difluorodichloromethane and adrenaline on different types of cardiac preparations isolated from sheep hearts. *Cardiovasc. Res.,* 20: 807-815.

LESSARD, Y., DESBROUSSES, S., & PAULET, G. (1977a) Arythmie cardiaque chez le lapin sous l'action de l'adrénaline et du difluorodichlorométhane (FC 12). *C.R. Soc. Biol. (Paris),* 171(4): 883-895.

LESSARD, Y., DESBROUSSES, S., & PAULET, G. (1977b) Arythmie cardiaque chez le chien sous l'action de l'adrénaline et du difluorodichlorométhane (FC 12). *C.R. Soc. Biol. (Paris),* 171(6): 1270-1282.

LESSARD, Y., DESBROUSSES, S., & PAULET, G. (1978) Rôle de l'adrénaline endogène dans le déclenchement de l'arythmie cardiaque par le difluorodichlorométhane (FC 12) chez les mammifères. *C.R. Soc. Biol. (Paris),* 172(2): 337-347.

LESSARD, Y., CALLEE, J.J., & PAULET, G. (1980) Action du difluorodichlorométhane (FC 12) sur l'activité électrique cardiaque du mammifere au niveau cellulaire. *C.R. Soc. Biol.,* 174(1): 52-57.

LESTER, D. & GREENBERG, L.A. (1950) Acute and chronic toxicity of some halogenated derivatives of methane and ethane. *Arch. ind. Hyg. occup. Med.,* 2: 335-344.

LEUSCHNER, F., NEUMANN, B.W., & HUEBSCHER, F. (1983) Report on subacute toxicological studies with several fluorocarbons in rats and dogs by inhalation. *Arzneimittelforschung,* 33(10): 1475-1476.

LEY, R.D. (1988) Induction of cutaneous melanomas by UV-B radiation in the marsupial *Monodelphis domestica.* In: *Proceedings of the 10th International Congress on Photobiology.*

LILLIAN, D., SINGH, H.B., APPLEBY, A., LOBBAN, L., ARNTS, R., GUMPERT, R., HAGUE, R., TOOMEY, J., KAZAZIS, J., ANTELL, M., HANSEN, D., & SCOTT, B. (1975) Atmospheric fates of halogenated compounds. *Environ. Sci. Technol.,* 9(12): 1042-1048.

LIPPMANN, M. (1989) Ozone health effects and emerging issues in relation to standards setting. In: Schneider, T., Lee, S.D., Grant, L.D., & Walters, G., ed. *Atmospheric ozone research and its policy implications: Third US-Dutch International Symposium, Nijmegen, The Netherlands, May 1988,* Amsterdam, Oxford, New York, Elsevier Science Publishers, pp. 21-33.

LONGSTAFF, E. & MCGREGOR, D.B. (1978) Mutagenicity of a halocarbon refrigerant monochlorodifluoromethane (R-22) in *Salmonella typhimurium. Toxicol. Lett.,* 2(1): 1-4.

LONGSTAFF, E., ROBINSON, M., BRADBROOK, C., STYLES, J.A., & PURCHASE, I.F.H. (1984) Genotoxicity and carcinogenicity of fluorocarbons: assessment by short-term *in vitro* tests and chronic exposure in rats. *Toxicol. appl. Pharmacol.,* 72(1): 15-31.

LOPRIENO, N., BARALE, R., PRESCIUTTINI, S., ROSSI, A.M., SBRANA, I., STRETTI, G., ZACCARO, L., ABBONDANDOLO, A., BONATTI, S., & FIORIO, R. (1979) Comparative data with different test systems using microorganisms and mammalian cells on references and environmental mutagens. *Mutat. Res.*, **64**: 119.

LOVELOCK, J.E. (1972) Atmospheric turbidity and trichlorofluoromethane. Concentrations in rural southern England and southern Ireland. *Atmos. Environ.*, **6**: 915.

LOVELOCK, J.E., MAGGS, R.J., & WADE, R.J. (1973) Halogenated hydrocarbons in and over the Atlantic. *Nature (Lond.)*, **241**: 194-196.

LOWENHEIM, F.A. & MORAN, M.K. (1975) *Faith, Keyes, and Clark's industrial chemicals*, 4th ed., New York, John Wiley and Sons, pp. 325-330.

LURE, A.Z. & PLESHKOVA, Z.I. (1977) [Comparative characteristics of the incidence of upper respiratory tract mucous membrane disease in workers of different shops of a fluoroorganic production plant.] *Zh. Ushn. Nos. Gorl. Bolezn.*, **3**: 47-50 (in Russian).

MCCARTHY, R.L. (1973) *Ecology and toxicology of fluorocarbons*, Wilmington, Delaware, E.I. Du Pont de Nemours & Co. (Unpublished report).

MCCAUL, J. (1971) Mix with care. *Environment*, **13**(1): 39.

MCCLURE, D.A. (1972) Failure of fluorocarbon propellants to alter the electrocardiogram of mice and dogs. *Toxicol. appl. Pharmacol.*, **22**: 221-230.

MCKNIGHT, J.E. & MCGRAW, J.L., Jr (1983) Ultrastructural study of the effects of trichlorotrifluoroethane on the liver of hairless mice. *J. submicrosc. Cytol.*, **15**(2): 447-451.

MAKIDE, Y., KANAI, Y., & TOMINGA, T. (1980) Background atmospheric concentrations of halogenated hydrocarbons in Japan. *Bull. Chem. Soc. Jpn*, **53**: 2681-2682.

MALTONI, C., CILIBERTI, A., & CARRETTI, D. (1982) Experimental contributions in identifying brain potential carcinogens in the petrochemical industry. *Ann. N.Y. Acad. Sci.*, **381**: 216-249.

MALTONI, C., LEFEMINE, G., TOVOLI, D., & PERINO, G. (1988) Long-term carcinogenicity bioassays on three chlorofluorocarbons (trichlorofluoromethane, FC-11; dichlorodifluoromethane, FC-12; chlorodifluoromethane, FC-22) administered by inhalation to Sprague-Dawley rats and Swiss mice. *Ann. N.Y. Acad. Sci.*, **534**: 261-282.

MARIER, G., MACFARLAND, G.H., & DUSSAULT, P. (1973) A study of blood fluorocarbon levels following exposure to a variety of household aerosols. *Household pers. Prod. Ind.*, **10**(12): 68, 70, 92, 99.

MATSUMOTO, T., PANI, K.C., KOVARIC, J.J., & HAMIT, H.F. (1968) Aerosol tissue adhesive spray: fate of freons and their acute topical and systemic toxicity. *Arch. Surg.*, 97(5): 727-735.

MAY, D.C. & BLOTZER, M.J. (1984) A report on occupation deaths attributed to Fluorocarbon-113. *Arch. environ. Health*, 39: 352-354.

MERGNER, G.W., BLAKE, D.A., & HELRICH, M. (1975) Biotransformation and elimination of ^{14}C-trichlorofluoromethane (FC-11) and ^{14}C-dichlorodifluoromethane (FC-12) in man (Laboratory Report). *Anaesthesiology*, 42(3): 345-351.

MORGAN, A., BLACK, A., WALSH, M., & BELCHER, D.R. (1972) The absorption and retention of inhaled fluorinated hydrocarbon vapors. *Int. J. appl. Radiat. Isot.*, 23: 285-291.

MORITA, M., MIKI, A., KAZAMA, H., & SAKATA, M. (1977) Case reports of deaths caused by freon gas. *Forensic Sci.*, 10(3): 253-260.

NASA (1986) *Present state of knowledge of the upper atmosphere: an assessment report*, Washington, DC, National Aeronautics and Space Administration (NASA Reference Publication No. 1162).

NASTAINCZYK, W., AHR, H.J., & ULLRICH, V. (1982a) The mechanism of the reproductive dehalogenation of polyhalogenated compounds by microsomal cytochrome P-450. *Adv. exp. Med. Biol.*, 136A: 799-808.

NASTAINCZYK, W., AHR, H.J., & ULLRICH, V. (1982b) The reductive metabolism of halogenated alkanes by liver microsomal cytochrome P-450. *Biochem. Pharmacol.*, 31(3): 391-396.

NCI (1978) *Bioassay of trichlorofluoromethane for possible carcinogenicity*, Bethesda, Maryland, National Cancer Institute.

NEELEY, W.B., BRANSON, D.R., & BLAU, G.E. (1974) Partition coefficient to measure bioconcentration potential of organic chemicals in fish. *Environ. Sci. Technol.*, 8(13): 1113-1115.

NIAZI, S. & CHIOU, W.L. (1975) Fluorocarbon aerosol propellants. IV. Pharmacokinetics of trichloromonofluoromethane following single and multiple dosing in dogs. *J. pharm. Sci.*, 64(5): 763-769.

NIAZI, S. & CHIOU, W.L. (1977) Fluorocarbon aerosol propellants. XI. Pharmacokinetics of dichlorodifluoromethane in dogs following single and multiple dosing. *J. pharm. Sci.*, 66: 49-53.

NIOSH (1980) *Health hazard evaluation determination*, Cincinnati, Ohio, US National Institute for Occupational Safety and Health (Report No. 79-127-644; NTIS PB 80-169816).

NOBLE, H.L. (1972) *Fluorocarbon industry review and forecast*, Chemical Marketing Research Association (Meeting May).

NRC (1976) *Environmental effects of chlorofluoromethane release*, Washington, DC, National Research Council, National Academy of Sciences, 125 pp.

NUCKOLLS, A.H. (1933) The comparative life, fire, and explosion hazards of common refrigerants, Chicago, Ohio, The Underwaters Laboratories, pp. 5-9, 24-26, 57-61, 106-107 (Miscellaneous Hazard No. 2375).

PARYJCZAK, T., BALCZEWSKA, H., IGNACZAK, W., KROL, A., & MAYTJEWSKI, P. (1985) [Determination of freons in air.] Bromatol. Chem. Toksykol., 18(1): 69-73 (in Polish).

PATERSON, J.W., SUDLOW, M.F., & WALKER, S.R. (1971) Blood levels of fluorinated hydrocarbons in asthmatic patients after inhalation of pressurized aerosols. Lancet, 2: 565-568.

PAULET, G. (1969) De l'action des hydrocarbures chlorofluorés sur l'organisme. Labo-Pharma. Probl. Tech., 180: 74-78.

PAULET, G. & CHEVRIER, R. (1969) Modalités de l'élimination par l'air expiré du fluorane 11 inhalé. Etude chez l'homme et chez l'animal. Arch. Mal. prof. Méd. Trav. Sécur. soc., 30(4-5): 251-256.

PAULET, G. & DESBROUSSES, S. (1969) Dichlorotetrafluoroethane. Acute and chronic toxicity. Arch. Mal. prof. Méd. Trav. Sécur. soc., 30: 477-492.

PAULET, G., DESBROUSSES, S., & VIDAL, E. (1974) Absence d'effet tératogène des fluorocarbones chez le rat et le lapin. Arch. Mal. prof. Méd. Trav. Sécur. soc., 35: 658.

PAULET, G., LANOE, J., THOS, A., TOULOUSE, P., & DASSONVILLE, J. (1975) Fate of fluorocarbons in the dog and rabbit after inhalation. Toxicol. appl. Pharmacol., 34(2): 204-213.

PITTS, D.G., ZUCLICH, J.A., ZIGMAN, S., ZIGLER, S., VARMA, S.D., MOSS, E., LERMAN, S., CAMERON, L.L., & JOSE, J. (in press) Optical radiation and cataracts. In: Waxhler, M. & Hitchins, V.M., ed. Optical radiation and visual health, Boca Raton, Florida, CRC Press.

PRENDERGAST, J.A., JONES, R.A., JENKINS, L.J., Jr, & SIEGEL, J. (1967) Effects on experimental animals of long-term inhalation of trichloroethylene, carbon tetrachloride, 1,1,1-trichloroethane, dichlorodifluoromethane, and 1,1-dichloroethylene. Toxicol. appl. Pharmacol., 10(2): 270-289.

PRINN, R.G., SIMMONDS, P.G., RASMUSSEN, R.A., ROSEN, R., ALYEA, F., CARDELINO, C., CRAWFORD, A., CUNNOLD, D., FRASER, P., & LOVELOCK, J. (1983) The atmospheric lifetime experiment. I. introduction, instrumentation, and overview. J. geophys. Res., 88(13): 8353-8367.

QUEVAUVILLER, A. (1965) Hygiène et sécurité des pulseurs pour aérosols médicamenteux. Prod. Probl. pharm., 20(1): 14-29.

QUEVAUVILLER, A., CHAIGNEAU, M., & SCHRENZEL, M. (1963) Etude expérimentale chez la souris de la tolérance du poumon aux hydrocarbures chlorofluorés. Ann. pharm. fr. 21(11): 727-734.

QUEVAUVILLER, A., SCHRENZEL, M., & VU-NGOC, H. (1964) Tolérance locale (peau, muqueuses, plaies, brûlures) chez l'animal aux hydrocarbures chlorofluorés. Thérapie, 19: 247-263.

References

RAFFI, G.B. & VIOLANTE, F.S. (1981) Is Freon 113 neurotoxic? A case report. *Int. Arch. occup. environ. Health*, **49**: 125-127.

RAMSEY, J.D. & FLANAGAN, R.J. (1982) Detection and identification of volatile organic compounds in blood by headspace gas chromatography as an aid to the diagnosis of solvent abuse. *J. Chromatogr.*, **240**(2): 423-444.

RAND (1986) *Product uses and market trends for potential ozone-depleting substances, 1985-2000*, Santa Monica, California, Rand Corporation (Report No. R-3386-EPA) (Prepared for the US Environmental Protection Agency).

RASMUSSEN, R.A. & KHALIL, M.A.K. (1982) Latitudinal distributions of trace gases in and above the boundary layer. *Chemosphere*, **11**(3): 227-235.

RASMUSSEN, R.A. & KHALIL, M.A.K. (1983) Natural and anthropogenic trace gases in the lower troposphere of the Arctic. *Chemosphere*, **12**(12): 371-375.

RASMUSSEN, R.A. & KHALIL, M.A.K. (1986) Atmospheric trace gases: Trends and distributions over the past decade. *Science*, **232**: 1623-1624.

RASMUSSEN, R.A., KHALIL, M.A.K., & DALLUGE, R.W. (1981) Atmospheric trace gases in Antarctica. *Science*, **211**(4479): 285-287.

RASMUSSEN, R.A., KHALIL, M.A.K., & CHANG, J.S. (1982) Atmospheric trace gases over China. *Environ. Sci. Technol.*, **16**(2): 124-126.

RAUWS, A.G., OLLING, M., & WIBOWO, A.E. (1973) Determination of fluorochlorocarbons in air and body fluids. *J. Pharm. Pharmacol.*, **25**: 718.

REBBERT, R.E. & AUSLOOS, P.J. (1975) Photodecomposition of $CFCl_3$ and CF_2Cl_2. *J. Photochem.*, **4**: 419-434.

REINEKE, F.J. & BAECHMANN, K. (1985) Gas chromatographic determination of C_2-C_8 hydrocarbons and halocarbons in ambient air by simultaneous use of three detectors. *J. Chromatogr.*, **323**(2): 323-329.

REINHARDT, C.F., MCLAUGHLIN, M., MAXFIELD, M.E., MULLIN, L.S., & SMITH, P.E., Jr (1971a) Human exposures to fluorocarbon 113. *Am. Ind. Hyg. Assoc. J.*, **32**: 143-152.

REINHARDT, C.F., AZAR, A., MAXFIELD, M.E., SMITH, P.E., Jr, & MULLIN, L.S. (1971b) Cardiac arrhythmias and aerosol "sniffing". *Arch. environ. Health*, **22**: 265-279.

REINHARDT, C.F., MULLIN, L.S., & MAXFIELD, M.E. (1973) Epinephrine-induced cardiac arrhythmia potential of some common industrial solvents. *J. occup. Med.*, **15**(12): 953-955.

ROBINSON, E., CRONN, D.R., MENZIA, F., CLARK, D., LEGG, R., & WATKINS, R. (1983) Trace gas profiles to 3000 m over Antarctica. *Atmos. Environ.*, **17**(5): 973-981.

RUDOLPH, J. & JEBSEN, C. (1983) The use of photoionization, flame ionization, and electron capture detectors in series for the determination of low molecular weight trace components in the non-urban atmosphere. *Int. J. environ. anal. Chem.*, 13(2): 129-139.

SAKATA, M., KAZAMA, H., MIKI, A., YOSHIDA, A., HAGA, M., & MORITA, M. (1981) Acute toxicity of fluorocarbon-22: toxic symptoms, lethal concentration, and its fate in rabbit and mouse. *Toxicol. appl. Pharmacol.*, 59(1): 64-70.

SALMON, A.G., JONES, R.B., & MACKRODT, W.C. (1981) Microsomal dechlorination of chloroethanes: structure-reactivity relationships. *Xenobiotica*, 11(11): 723-734.

SALMON, A.G., NASH, J.A., WALKLIN, C.M., & FREEDMAN, R.B. (1985) Dechlorination of halocarbons by microsomes and vesicular reconstituted cytochrome P-450 systems under reductive conditions. *Br. J. ind. Med.*, 42(5): 305-311.

SALTZMAN, B.E., COLEMAN, A.I., & CLEMONS, C.A. (1966) Halogenated compounds as gaseous meteorological tracers: stability and ultra-sensitive analysis by gas chromatography. *Anal. Chem.*, 38: 753-758.

SANDERS, P.A. (1960) Corrosion of aerosol cans. *Soap Chem. Spec.*, 36: 95-103.

SAYERS, R.R., YANT, W.P., CHORNYAK, J., & SHOAF, H.W. (1930) *Toxicity of dichlorofluoromethane: A new refrigerant*, Washington, DC, Bureau of Mines, Publications Distribution Section (Report of Investigations No. 3013).

SCHNEIDER, T., LEE, S.D., GRANT, L.D., & WOLTERS, G., ed. (1989) *Atmospheric ozone research and its policy implications: Third US-Dutch International Symposium, Nijmegen, The Netherlands, May 1988*, Amsterdam, Oxford, New York, Elsevier Science Publishers, 1047 pp.

SCHOLZ, J. (1961) [*Progress in biological aerosol research*], Stuttgart, F.K. Schattaver Verlag (in German).

SCHOLZ, J. (1962) [New toxicological investigations on certain types of freon used as propellants.] *Fortschr. Biol. Aerosol-Forsch.*, 4: 420-429 (in German).

SETLOW, R.B. (1988) A fish model for the induction of melanomas by UV-B radiation. In: *Proceedings of the 10th International Congress of Photobiology, Tel Aviv, Israel, 27 October - 4 November, 1988*, Washington, DC, Department of Energy (Foreign Trip Report NTIS/DE 89008098).

SHARGEL, L. & KOSS, R. (1972) Determination of fluorinated hydrocarbon propellants in the blood of dogs after aerosol administration. *J. pharm. Sci.*, 61(9): 1445-1449.

SHERMAN, H. (1974) *Long-term feeding studies in rats and dogs with dichlorodifluoromethane (freon 12 food freezant)*, Newark, Delaware, Haskell Laboratory (Medical Research Project No. 1388; Report No. 24-74) (Unpublished, courtesy of Du Pont de Nemours Co.).

SILVERGLADE, A. (1971) Evaluation of reports of death from asthma. *J. asthma Res.*, **8**: 95.

SINGH, H.B., SALAS, L.J., SHIGEISHI, H., & SCRIBNER, E. (1979) Atmospheric halocarbons, hydrocarbons, and sulfur hexafluoride: global distributions, sources, and sinks. *Science*, **203**: 899-903.

SINGH, H.B., SALAS, L.J., SMITH, A.J., & SHIGEISHI, H. (1981) Measurements of some potentially hazardous organic chemicals in urban environments. *Atmos. Environ.*, **15**: 601-612.

SINGH, H.B., SALAS, L.J., & STILES, R.E. (1983) Selected man-made halogenated chemicals in the air and oceanic environment. *J. geophys. Res.*, **88**(C6): 3675-3683.

SKURIC, Z., ZUSKIN, E., & VALIC, F. (1975) Effects of aerosols in common use on the ventilatory capacity of the lung. *Int. Arch. Arbeitsmed.*, **34**: 137-149.

SLAPER, H. (1987) *Skin cancer and UV exposure: Investigations on the estimation of risks*, Utrecht, Federal Republic of Germany, University of Utrecht (Ph. D. Thesis).

SLATER, T.E. (1965) A note on the relative toxic activities of tetrachloromethane and trichlorofluoromethane on the rat. *Biochem. Pharmacol.*, **14**(2): 178-181.

SMART, B.E. (1980) Fluorinated aliphatic compounds. In: Grayson, M. & Eckroth, D., ed. *Kirk-Othmer encyclopedia of chemical technology*, 3rd ed., New York, John Wiley and Sons, Vol. 10, pp. 856-870.

SMITH, J.K. & CASE, M.T. (1973) Subacute and chronic toxicity studies of fluorocarbon propellants in mice, rats, and dogs. *Toxicol. appl. Pharmacol.*, **26**(3): 438-443.

SPEIZER, F.E., WEGMAN, D.H., & RAMIREZ, A. (1975) Palpitation rates associated with fluorocarbon exposure in hospital setting. *New Engl. J. Med.*, **292**: 624.

SRI (1986) *1986 Directory of chemical producers, United States of America*, Menlo Park, California, Stanford Research Institute International, pp. 550, 554, 557, 586, 588, 1067.

STEINBERG, M.B., BOLDT, R.E., RENNE, R.A., & WEEKS, M.H. (1969) *Inhalation toxicity of 1,1,2-trichloro-1,2,2-trifluoroethane (TCTFE)*, Edgewood Arsenal, Maryland, US Army Environmental Hygiene Agency (NTIS AD Report No. 854-705).

STEWART, R.D., NEWTON, P.E., BARETTA, E.D., HERRMANN, A.A., FORST, H.V., & SOTO, R.J. (1978) Physiological response to aerosol propellants. *Environ. Health Perspect.*, 26: 275-285.

STOPPS, G.J. & MCLAUGHLIN, M. (1967) Psychophysiological testing of human subjects exposed to solvent vapors. *Am. Ind. Hyg. Assoc. J.*, 28: 43-50.

SU, C.W. & GOLDBERG, E.D. (1973) Chlorofluorocarbons in the atmosphere. *Nature (Lond.)*, 245: 27.

SU, C.W. & GOLDBERG, E.D. (1976) Environmental concentrations and fluxes of some halocarbons. In: Windom, H.L. & Due, R.A., ed. *Marine pollutant transfer*, Lexington, Massachusetts, Lexington Books, DC, Health and Company, pp. 353-374.

SULLIVAN, J. & TERAMURA, A.H. (1988) Effects of UV-B radiation on seedling growth in the Pinaceae. *Am. J. Bot.*, 75: 225-230.

SWIFT, D.L., ZUSKIN, E., & BOUHUYS, A. (1979) Respiratory deposition of hair spray aerosol and acute lung function changes. *Lung*, 156: 149-158.

SZMIDT, M., AXELSON, O., & EDLING, C. (1981) [Cohort study of Freon-exposed workers.] *Acta Soc. Med. Svec. Hygiea*, 90: 77-79 (in Swedish).

TAYLOR, G.J. & DREW, R.T. (1974) Cardiovascular effects of acute and chronic inhalations of fluorocarbon 12 in rabbits. *J. Pharmacol. exp. Ther.*, 192(1): 129-135.

TAYLOR, G.J. & HARRIS, W.S. (1970a) Cardiac toxicity of aerosol propellants. *J. Am. Med. Assoc.*, 214: 81.

TAYLOR, G.J. & HARRIS, W.S. (1970b) Glue sniffing causes heart block in mice. *Science*, 170: 866.

TAYLOR, G.J., IV, HARRIS, W.S., & BOGDONOFF, M.D. (1971) Ventricular arrhythmias induced in monkeys by the inhalation of aerosol propellants. *J. clin. Invest.*, 50: 1546-1550.

TERAMURA, A.H. (1983) Effects of ultraviolet-B radiation on the growth and yield of crop plants. *Physiol. Plant.*, 58: 415-427.

TERAMURA, A.H. & SULLIVAN, J. (1988) *Annual report to the US Environmental Protection Agency: The effects of changing climate and stratospheric ozone modification on plants*, College Park, Maryland, University of Maryland.

TEVINI, M. & IWANZIK, W. (1986) Effects of UV-B radiation on growth and development of cucumber seedlings. In: Worrest, R.C. & Caldwell, M.M., ed. *Stratospheric ozone reduction, solar UV radiation and plant life*, Heidelberg, Springer-Verlag, pp. 271-285.

TEVINI, M., STEINMULLER, D., & IWANZIK, W. (1986) [*The effect of UV-B radiation, in combination with other stress factors, on the growth and function of useful plants*], Munich, GSF (Report 6/86) (in German).

THOMAS, G. (1965) Narcotic effects of acute exposure to trichlorofluoromethane (Freon 11). *Trans. Assoc. Ind. Med. Off.*, **15**(3): 105-106.

TOMATIS, L., PARTENSKY, C., & MONTESANO, R. (1973) The predictive value of mouse liver tumor induction in carcinogenicity testing: a literature survey. *Int. J. Cancer*, **12**: 1-20.

TOMITA, I., SAITOU, S., MEGURO, M., & KANAMORI, H. (1983) [Studies on the photolysis of trichlorofluoromethane and dichlorodifluoromethane in water. III. Determination and dichlorodifluoromethane in coastal water.] *Eisei Kagaku*, **29**(2): 76-82 (in Japanese).

TROCHIMOWICZ, J. (1984) *The toxicology of Fluorocarbon 113*. Presented to the Swedish Board of Occupational Safety and Health, Stockholm, 19 January 1984 (Prepared by Haskell Laboratory for Toxicology and Industrial Medicine for E.J. du Pont de Nemours & Co.).

TROCHIMOWICZ, H.J., AZAR, A., TERRILL, J.B., & MULLIN, L.S. (1974) Blood levels of fluorocarbon related to cardiac sensitization: Part II. *J. Am. Ind. Hyg. Assoc.*, **35**: 632.

TROCHIMOWICZ, H.J., RUSCH, G.M., CHIU, T., & WOOD, C.K. (1988) Chronic inhalation toxicity/carcinogenicity study in rats exposed to fluorocarbon 113 (FC-113). *Fundam. appl. Toxicol.*, **11**: 68-75.

TRUHAUT, R., BOUDENE, C., JOUANY, J.M., & BOUANT, A. (1972) Experimental study of the toxicity of a fluoroalkene derivative, hexafluorodichlorobutene (HFCB). *Fluoride*, **5**: 4.

TYRAS, H. (1981) Spectrophotometric determination of trichlorofluoromethane in air. *Fresenius Z. anal. Chem.*, **309**(5): 400.

TYSON, B.J., ARVESEN, J.C., & OHARA, D. (1978) Interhemispheric gradients of CF_2Cl_2, $CFCl_3$, CCl_4, and N_2O. *Geophys. Res. Lett.*, **5**: 535-538.

UEHLEKE, H., WERNER, T., GREIM, H., & KRAMER, M. (1977) Metabolic activation of haloalkanes and tests *in vitro* for mutagenicity. *Xenobiotica*, **7**(7): 393-400.

URBACH, F., ed. (1969) *The biological effects of ultraviolet radiation: With emphasis on skin*, Oxford, New York, Pergamon Press.

US EPA (1975) *Report on the problem of halogenated air pollutants and stratospheric ozone*, Research Triangle Park, North Carolina, US Environmental Protection Agency, Office of Research and Development, 55 pp (EPA-600/9-75-008).

US EPA (1978) Toxic substance control act clarification of 43 FR 11318: Ban, effective date 12/15/78. *Fed. Reg.*, **43**: 55241.

US EPA (1980) *Computer printout of non-confidential production data from TSCA inventory*, Washington, DC, US Environmental Protection Agency, Office of Pesticides and Toxic Substances, Chemical Information Division.

US EPA (1983) *Health assessment document for 1,1,2-trichloro-1,2,2-trifluoroethane (chlorofluorocarbon CFC-113),* Cincinnati, Ohio, US Environmental Protection Agency, Office of Health and Environmental Assessment, Environmental Criteria and Assessment Office (EPA-600/8-82-002F; NTIS PB 84-118843).

US EPA (1987a) *Assessing the risks of trace gases that can modify the stratosphere: v. I-V,* Washington, DC, US Environmental Protection Agency, Office of Air and Radiation (EPA Report No. 400/1-87/001A-E).

US EPA (1987b) Risks to crops and terrestrial ecosystems from enhanced UV-B radiation. In: Hoffman, J., ed. *Assessing the risks of trace gases that can modify the stratosphere,* Washington, DC, US Environmental Protection Agency, Office of Air and Radiation, pp (11)1-31 (EPA Report No. 400/1-87/001C).

US EPA (1987c) An assessment of the effects of ultraviolet-B radiation on aquatic organisms. In: Hoffman, J., ed. *Assessing the risk of trace gases that can modify the stratosphere,* Washington, DC, US Environmental Protection Agency, Office of Air and Radiation, pp. (12)1-33 (EPA Report No. 400/1-87/001C).

US EPA (1987d) *Ultraviolet radiation and melanoma: With special focus on assessing the risks of stratospheric ozone depletion,* Washington, DC, US Environmental Protection Agency.

US ITC (1975) *Synthetic organic chemical: United States production and sales, 1973,* Washington, DC, US International Trade Commission, p. 205 (US ITC Publication No. 728).

US ITC (1976) *Synthetic organic chemical: United States production and sales, 1974,* Washington, DC, US International Trade Commission, p. 203 (US ITC Publication No. 776).

US ITC (1977a) *Synthetic organic chemical: United States production and sales, 1976,* Washington, DC, US International Trade Commission, p. 302 (US ITC Publication No. 833).

US ITC (1977b) *Synthetic organic chemical: United States production and sales, 1975,* Washington, DC, US International Trade Commission, p. 198 (US ITC Publication No. 804).

US ITC (1978) *Synthetic organic chemical: United States production and sales, 1977,* Washington, DC, US International Trade Commission, p. 920 (US ITC Publication No. 920).

US ITC (1979) *Synthetic organic chemical: United States production and sales, 1978,* Washington, DC, US International Trade Commission, p. 313 (US ITC Publication No. 1001).

US ITC (1980) *Synthetic organic chemical: United States production and sales, 1979,* Washington, DC, US International Trade Commission, p. 267 (US ITC Publication No. 1099).

US ITC (1981) *Synthetic organic chemical: United States production and sales, 1980,* Washington, DC, US International Trade Commission, pp. 162, 265, 298 (US ITC Publication No. 1183).

US ITC (1982) *Synthetic organic chemical: United States production and sales, 1981,* Washington, DC, US International Trade Commission, p. 261 (US ITC Publication No. 1422).

US ITC (1985) *Synthetic organic chemical: United States production and sales, 1984,* Washington, DC, US International Trade Commission, p. 258 (US ITC Publication No. 1745).

VAINIO, H., NICKELS, J., & HEINONEN, T. (1980) Dose-related hepatotoxicity of 1,1,2-trichloro-1,2,2-trifluoroethane in short-term intermittent inhalation exposure in rats. *Toxicology,* 18(1): 17-35.

VALIC, F., ZUSKIN, E., SKURIC, Z., & DENICH, M. (1974) [Effects of aerosols of common use on the ventilatory lung capacity. 1. Change in $FEV_{1.0}$ in exposure to hair sprays.] *Acta med. Yugosl.,* 28: 231-246 (in Croatian with English summary).

VALIC, F., SKURIC, Z., BANTIC, Z., RUDAR, M., & HECEJ, M. (1977) Effects of fluorocarbon propellants on respiratory flow and ECG. *Br. J. ind. Med.,* 34: 130-136.

VALIC, F., SKURIC, Z., & ZUSKIN, E. (1982) [Experimental exposure to freons 12, 22, and 502.] *Rad Jazu,* 402(18): 229-243 (in Croatian with English Summary).

VAN AUKEN, O.W. & WILSON, R.H. (1973) Halogenated hydrocarbon induced reduction in coupling parameters of rabbit liver and mung bean mitochondria. *Naturwissenschaften,* 60: 259.

VAN DER LEUN, J.C. (1988) *Effects of ozone depletion: UNEP scientific review of ozone layer modification and its impact,* Nairobi, United Nations Environment Programme (UNEP/OZ.L.Sc.1/5).

VAN DER LEUN, J.C. (1989) Effects of increased UV-B on human health. In: Schneider, T., Lee, S.D., Grant, L.D., & Wolters, G., ed. *Atmospheric ozone research and its policy implications.: Third US-Dutch International Symposium, Nijmegen, The Netherlands, May 1988,* Amsterdam, Oxford, New York, Elsevier Science publishers, pp. 803-812.

VAN KETEL, W.G. (1976) Allergic contact dermatitis from propellants in deodorant sprays in combination with ethyl chloride. *Contact dermatitis,* 2: 115-119.

VAN STEE, E.W. & MCCONNELL, E.E. (1977) Studies of the effects of chronic inhalation exposure of rabbits to chlorodifluoromethane. *Environ. Health Perspect.,* 20: 246-247.

VAN'T HOF, J. & SCHAIRER, L.A. (1982) Tradescantia assay system for gaseous mutagens: A report of the US Environmental Protection Agency gene-tox program. *Mutat. Res.,* 99(3): 303-315.

WARD, R. (1983) *E.I. Du Pont de Nemours and Company: transmittal of summary of teratogenicity study of 1,1,2-trichloro-1,2,2-trifluoroethane in rats*, Alderly Park, Macclesfield, Imperial Chemical Industries PLC, Central Toxicology Laboratory (Report No. CTL/P/731).

WARITZ, R.S. (1971) *Toxicology of some commercial fluorocarbons*, Springfield, Virginia, National Technical Information Service (NTIS AD Report No. 751429).

WATANABE, T. & AVIADO, D.M. (1975) Toxicity of aerosol propellants in the respiratory and circulatory systems. IV. Influence of pulmonary emphysema and anesthesia in the rat. *Toxicology*, 3: 225-236.

WEIGAND, W., von (1971) [Investigations on the inhalation toxicity of fluorine derivatives of methane, ethane and cyclobutane.] *Zentralbl. Arbeitsmed. Arbeitsschutz*, 21(5): 149-156 (in German).

WILLS, J.H. (1972) Sensitization of the heart to catecholamine-induced arrhythmia. In: *Proceedings of the 3rd Annual Conference on Environmental Toxicology, Fairborn, Ohio, 25-27 October* Springfield, Virginia, National Technical Information Service, p. 249 (NTIS AD Report No. 773766).

WINDHOLZ, M., ed. (1983) *The Merck index*, 10th ed., Rahway, New Jersey, Merck and Company.

WMO/CANADA DOE (1989) *Proceedings of the World Conference on the Changing Atmosphere, Toronto, Canada, June 1988*, Geneva, World Meteorology Organization and Ottawa, Canada Department of the Environment, 483 pp.

WOLF, C.R., MANSUY, D., NASTAINCZYK, W., DEUTSCHMANN, G., & ULLRICH, V. (1977) The reduction of polyhalogenated methanes by liver microsomal cytochrome P-450. *Mol. Pharmacol.*, 13(4): 698-705.

WOLF, C.R., KING, L.J., & PARKE, D.V. (1978) The anaerobic dechlorination of trichlorofluoromethane by rat liver preparations *in vitro*. *Chem.-biol. Interact.*, 21(2-3): 277-288.

WOOD, C.K. (1985) *Two-year inhalation toxicity study with 1,1,2-trichloro-1,2,2-trifluoroethane in rats*, Newark, Delaware, Haskell Laboratory, 63 pp (Medical Research Project No. 3683-001; Report No. 488-84) (Unpublished, courtesy of Du Pont de Nemours & Co.).

WORREST, R.C. (1982) Review of literature concerning the impact of UV-B radiation upon marine organisms. In: Calkins, J., ed. *The role of solar ultraviolet radiation in marine ecosystems*, New York, London, Plenum Press, pp. 429-458.

WORREST, R.C. (1983) Impact of solar ultraviolet-B (290-320 nm) upon marine microalgae. *Physiol. Plant.*, 58: 428-434.

WORREST, R.C., VAN DYKE, H., & THOMSON, B.E. (1978) Impact of enhanced simulated solar ultraviolet radiation upon a marine community. *Photochem. Photobiol.*, 27: 471-478.

WORREST, R.C., THOMSON, B.E., & VAN DYKE, H. (1981a) Impact of UV-B radiation upon estuarine microcosms. *Photochem. Photobiol.*, 33: 223-227.

WORREST, R.C., WOLNIAKOWSKI, K.U., SCOTT, J.D., BROOKER, D.L., THOMSON, B.E., & VAN DYKE, H. (1981b) Sensitivity of marine phytoplankton to UV-B radiation (290-320 nm): impact upon a model ecosystem. *Photochem. Photobiol.*, **33**: 223-227.

YANT, W.P., SCHRENK, H.H., & PATTY, F.A. (1932) *Toxicity of dichlorotetrafluoroethane*, Washington, DC, US Bureau of Mines (Report No. R.I. 3185).

YOUNG, W. & PARKER, J.A. (1972) Effect of freons on acetylcholinesterase activity and some counter measures. In: *Proceedings of the 3rd Annual Conference on Environmental Toxicology, Fairborn, Ohio, 25-27 October*, Springfield, Virginia, National Technical Information Service, p. 259 (NTIS AD Report No. 773766).

ZAKHARI, S. & AVIADO, D.M. (1982) Cardiovascular toxicology of aerosol propellants, refrigerants, and related solvents. In: Van Stee, E.W., ed. *Cardiovascular toxicology*, New York, Raven Press, pp. 281-314.

ZASAVITSKII, I.I., KOSICHKIN, Y.V., NADEZHDINSKII, A.I., STEPANOV, E., TISHCHENKO, A., KHATTATOV, V., & SHOTOV, A. (1984) [Study of absorption spectra for the detection of small dichlorodifluoromethane concentrations using a diode laser.] *Zh. Prikl. Spektrosk.*, **41**(3): 396-401 (in Russian).

ZURER, P.S. (1988) Search intensifies for alternatives to ozone-depleting halocarbons. *Chem. eng. News*, 8 February: 17-20.

ZUSKIN, E. & BOUHUYS, A. (1974) Acute airway responses to hair-spray preparations. *N. Engl. J. Med.*, **290**: 660-663.

ZUSKIN, E., SKURIC, Z., & VALIC, F. (1974) [Effects of aerosols of common use on the ventilatory lung capacity. II.] *Acta med. Yugosl.*, **28**: 247-259 (in Croatian with English summary).

ZUSKIN, E., LOKE, J., & BOUHUYS, A. (1981) Helim-oxygen flow-volume curves in detecting acute response to hair spray. *Int. Arch occup. environ. Health*, **49**: 41-44.

RESUME

1. Identité, propriétés physiques et chimiques, méthodes d'analyse

La présente monographie ne traite que des chlorofluorocarbures (CFC) obtenus par substitution de la totalité des atomes d'hydrogène du méthane et de l'éthane par des atomes de fluor et de chlore. Nombre de ces composés ont un intérêt commercial et certains d'entre eux contribuent à la diminution de la couche d'ozone. Les composés examinés dans ce qui suit sont les suivants: le trichlorofluorométhane (CFC-11), le dichlorofluorométhane (CFC-12), le chlorotrifluorométhane (CFC-13), le difluoro-1,2-tétrachloro-1,1,2,2-éthane (CFC-112), le difluoro-1,1-tétrachloro-1,2,2,2-éthane (CFC-112a), le trichloro-1,1,2-trifluoro-1,2,2-éthane (CFC-113), le trichloro-1,1,1-trifluoro-2,2,2-éthane (CFC-113a), le dichloro-1,1,2,2-tétrafluoro-1,1,2,2-éthane (CFC-114), le dichloro-1,1-tétrafluoro-1,2,2,2-éthane (CFC-114a) et le chloro-1-pentafluoro-1,1,2,2,2-éthane (CFC-115). Les composés qui ne contiennent pas de chlore (comme le CFC-134a et le CFC-116) ne sont pas examinés. Quant aux composés qui contiennent de l'hydrogène (comme le chlorodifluorométhane), ils feront l'objet d'un rapport ultérieur.

Les chlorofluorocarbures du commerce comptent parmi les produits organiques les plus purs qu'on puisse obtenir. Ils se caractérisent en général par une tension de vapeur et une densité élevées ainsi que par une viscosité, une tension superficielle, un indice de réfraction et une solubilité dans l'eau faibles. Le degré de substitution par le fluor modifie fortement les propriétés physiques et en général à mesure qu'il augmente, la tension de vapeur augmente et le point d'ébulition, la densité et la solubilité dans l'eau diminuent.

Les chlorofluorocarbures examinés dans la présente monographie sont assez stables chimiquement et, en l'absence de catalyseur métallique, leur vitesse d'hydrolyse est faible. Ils sont très résistants aux oxydants classiques aux températures inférieures à 200 °C. En général, les chlorofluorocarbures présentent une grande stabilité thermique et sont extrêmement résistants à la presque totalité des agents chimiques. Toutefois ils réagissent violemment sur les métaux de forte réactivité chimique.

Plusieurs méthodes d'analyse sont utilisables pour le dosage des chlorofluorocarbures dans divers milieux. Il s'agit de la spectrophotométrie, de la chromatographie en phase gazeuse avec plusieurs variantes et de la spectrométrie de masse. Dans la plupart des cas on utilise la chromatographie en phase gazeuse avec différents modes de détection; dans ce cas, les limites de détection sont généralement de l'ordre d'une partie par trillion (ppt). Les méthodes de prélèvement des échantillons ont été modifiées pour obtenir une sélectivité et une sensibilité accrues.

2. Sources d'exposition humaine et environnementales

Les chlorofluorocarbures étudiés dans la présente monographie n'existent pas à l'état naturel mais presque tous, sauf ceux qui sont utilisés comme intermédiaires chimiques, sont libérés dans l'environnement. En 1985, on estime que la production mondiale des chlorofluorocarbures les plus importants susceptibles de provoquer une diminution de la couche d'ozone (CFC-11, CFC-12, CFC-113) était d'au moins un million de tonnes. La production n'est pas limitée aux grands pays industriels puisqu'au moins 16 pays en fabriquent. La mise en oeuvre du Protocole de Montréal va probablement provoquer un renversement de la tendance actuelle qui est à l'accroissement de la production.

La méthode la plus importante pour la préparation des principaux chlorofluorocarbures consiste dans le déplacement catalytique du chlore des chlorocarbures par le fluor en présence de gaz fluorhydrique anhydre. La plupart des émissions dans l'environnement se produisent lors de la mise au rebut de matériel de réfrigération plutôt que pendant la fabrication, le stockage ou la manipulation de ces produits. Les restrictions imposées par la loi à l'utilisation de ces produits dans de nombreux pays ont réduit la libération dans l'atmosphère des chlorofluorocarbures utilisés comme gaz propulseurs; quant aux émissions d'agents de soufflage, elles sont faibles. En raison de la forte tension de vapeur de ces produits à la température ambiante, presque toute la masse libérée dans l'environnement finit par s'accumuler dans l'atmosphère. On estime qu'en 1985 les émissions annuelles totales de chlorofluorocarbures, principalement du CFC-11 et du CFC-12, atteignaient environ un million de tonnes, les émissions cumulées de ces produits entre 1931 et 1985 s'élevant à environ 13,5 millions de tonnes.

En 1985, les différentes utilisations dans le monde des chlorofluorocarbures se décomposaient comme suit: réfrigérants 15%, agents de soufflage 35%, gaz propulseurs pour aérosols 31%, divers 7%, sans utilisation définie 12%. Aux Etats-Unis d'Amérique, la proportion de chlorofluorocarbures utilisés comme gaz propulseurs était beaucoup plus faible en raison des restrictions imposées à leur utilisation.

3. Transport, distibution et transformation dans l'environnment

Les chlorofluorocarbures du commerce persistent dans l'environnement en raison de leur stabilité chimique. La durée moyenne de persistance dans l'atmosphère est estimée à 65, 110, 400, 90, 180 et 380 ans pour le CFC-11, le CFC-12, le CFC-13, le CFC-113, le CFC-114 et le CFC-115 respectivement. Cette longue persistance permet une diffusion dans la stratosphère où les chlorofluorocarbures donnent naissance par voie photochimique à des atomes de chlore qui réagissent sur la couche d'ozone. En outre, ces composés contribuent à l'effet de serre.

4. Niveaux dans l'environnement et exposition humaine

Plusieurs chercheurs ont fait état de la distribution mondiale des chlorofluorocarbures. Des mesures récentes portant sur la variation en fonction de la latitude de la concentration en chlorofluorocarbures ont montré qu'il n'y a guère de différence entre l'hémisphère nord et l'hémisphère sud pour ce qui concerne le CFC-11 et le CFC-12. Il n'y a également guère de variations en fonction de l'altitude jusqu'à 6 km au-dessus de la surface terrestre. Les concentrations mesurées dans l'air des villes et des banlieues sont plus élevées que celles qu'on observe dans les régions rurales ou écartées en raison d'émissions locales.

Résumé

Les concentrations atmosphériques de CFC-11 et CFC-12 ont augmenté régulièrement jusqu'en 1985 où leurs concentrations globales aux Etats-Unis d'Amérique atteignaient 9120 ng/m³ dans les zones urbaines et suburbaines, et 2720 ng/m³ dans les zones rurales ou écartées. A partir de ces données, on estime que l'exposition humaine par inhalation dans ces deux types de secteur s'établissait respectivement à 182 et 54 mg/jour.

A la surface des océans, la concentration moyenne de CFC-11 et de CFC-12 mesurée dans trois secteurs éloignés s'est révélée de l'ordre de 0,2 ng/litre. Toutefois on a mesuré des concentrations de 0,62 ng de CFC-11/litre dans la Mer du Groenland en 1982 et des valeurs atteignant 0,54 ng/litre ont été observées dans les eaux côtières du Japon. En ce qui concerne le CFC-12, la plus forte concentration dans ces mêmes eaux était de 0,33 ng/litre. Des valeurs beaucoup plus élevées ont été obtenues dans les eaux douces du Lac Ontario où on a mesuré des concentrations de 249 mg de CFC-11 par litre et de 572 ng de CFC-12 par litre. On n'a pas décelé de chlorofluorocarbures dans l'eau de boisson mais ils étaient présents dans la neige et dans l'eau de pluie en Alaska, dans le Lac Ontario et dans le Niagara. Du CFC-11 a été décelé à des concentrations de 0,1 à 5 μg/kg de poids à sec (parties par milliard) dans divers organes de poissons et de mollusques. Toutefois la présence de chlorofluorocarbures dans les préparations alimentaires n'est pas attestée.

5. Cinétique et métabolisme

Des chlorofluorocarbures peuvent pénétrer dans l'organisme humain par inhalation, ingestion ou contact cutané. C'est l'inhalation qui constitue la voie de pénétration la plus fréquente et la plus importante, la voie d'élimination principale étant l'air exhalé. Des études contrôlées sur des volontaires et des animaux d'expérience ont fourni des données substantielles relatives à l'exposition à un certain nombre de chlorofluorocarbures. Ces données montrent que les chlorofluorocarbures:

- peuvent être absorbés au niveau de la membrane alvéolaire, des voies digestives ou de la peau;
- passent rapidement dans le sang après inhalation;

- passent dans le sang d'autant moins vite que leur concentration y est plus élevée;
- une fois dans le sang, sont absorbés par divers tissus;
- finissent par atteindre une concentration sanguine stationnaire au bout d'une durée d'exposition suffisante, ce qui indique qu'un équilibre s'établit entre l'air chargé de chlorofluorocarbures et le sang;
- continuent à être absorbés par les tissus après stabilisation du taux sanguin initial et continuent de pénétrer dans l'organisme.

L'expérimentation animale montre que les chlorofluorocarbures sont rapidement absorbés après inhalation et sont amenés par le sang à la presque totalité des tissus. C'est dans les tissus adipeux ou lipidiques qu'ils atteignent en général les concentrations les plus élevées. Toutefois on trouve également des chlorofluorocarbures dans les organes bien irrigués comme le coeur, les poumons, les reins et les muscles.

Les résultats fournis par des études métaboliques sur l'animal et sur l'homme montrent que les chlorofluorocarbures résistent à la dégradation ou à la métabolisation par les systèmes biologiques. En fait, après exposition, les chlorofluorocarbures sont généralement très peu ou pas du tout métabolisés.

Quelle que soit la voie de pénétration, les chlorofluorocarbures sont éliminés presque exclusivement dans l'air expiré. Lors d'études visant à rechercher d'éventuels produits de transformation métabolique dans les urines ou les matières fécales, on n'a pas récupéré ces produits ou leurs métabolites en quantités notables.

6. Effets sur l'environnement

Certains chlorofluorocarbures, notamment les CFC-11, 12, 113, 114 et 115 sont extrêmement stables dans les conditions qui règnent dans la basse atmosphère. Ce n'est qu'après être parvenus dans la haute atmosphère qu'ils perdent des atomes de chlore par photolyse sous l'influence des rayonnements de haute énergie qui y pénètrent.

Ces radicaux chlore provoquent la destruction catalytique de l'ozone. L'ozone stratosphérique absorbe le rayonnement ultra-violet solaire (UV-B de 280-320 nm de longueur d'onde), de sorte que seule une partie de ce rayonnement parvient jusqu'à la surface de la terre.

L'expérience montre qu'un acroissement du rayonnement UV-B à la surface de la terre par suite de la diminution de la couche d'ozone aurait des effets délétères sur les organismes terrestres et aquatiques. Malgré les incertitudes tenant à la complexité de l'expérimentation sur le terrain, les données disponibles montrent que le rendement des récoltes et la productivité des forêts auraient à souffrir d'un accroissement du rayonnement UV-B solaire. Il semble également que l'accroissement de ce rayonnement puisse modifier la répartition et l'abondance des végétaux ainsi que la structure de l'écosystème.

Différentes études portant sur des écosystèmes marins ont montré que le rayonnement UV-B pouvait être nocif pour les larves de poissons et les alevins, les larves de crevettes et de crabes, les copépodes ainsi que les végétaux qui sont essentiels au réseau alimentaire marin. Parmi ces effets nocifs, on peut citer une réduction de la fécondité, de la croissance et de la survie. L'expérience montre que même une faible augmentation de l'exposition au rayonnement UV-B ambiant pourrait entraîner une modification sensible de l'écosystème.

7. Effets sur les animaux d'expérience et les systèmes in vitro

On a largement étudié la toxicité aiguë des chlorofluorocarbures après inhalation. Ceux qui sont envisagés dans la présente monographie sont faiblement toxiques lorsqu'ils sont inhalés. Les symptômes d'une intoxication aiguë comportent des effets sur le système nerveux central, des effets cardiovasculaires secondaires ainsi qu'une irritation des voies respiratoires. Le peu de données dont on dispose au sujet de la toxicité aiguë par voie orale des chlorofluorocarbures montrent que celle-ci est faible. Appliqués sur la peau à hautes doses, le CFC-112, le CFC-112a et le CFC-113 provoquent une irritation d'intensité variable mais pas d'autres effets sensibles.

Des études d'inhalation de courte durée ont été effectuées sur le CFC-11, le CFC-12, le CFC-112, le CFC-113, le CFC-114 et le CFC-115. Les résultats indiquent que la toxicité est faible, les effets observés s'exerçant principalement au niveau du système nerveux central, des voies respiratoires et du foie. Les études de toxicité par voie orale confirment la faible toxicité de ces produits.

Lors d'une étude d'inhalation à long terme, des rats ont été exposés à du CFC-113 aux doses de 0,2, 1 ou 2% (15,3, 76,6 ou 183 g/m^3) six heures par jour, cinq jours par semaine pendant des périodes allant jusqu'à deux ans. Aucune anomalie n'a été observée, ni sur le plan histologique ni en ce qui concerne les résultats des analyses biologiques. La seule anomalie que les auteurs ont imputée à ce traitement était une réduction du gain de poids corporel chez les groupes exposés aux deux doses les plus élevées.

Les données disponibles indiquent que les chlorofluorocarbures complètement halogénés examinés dans la présente monographie ont une activité mutagène ou cancérogène faible ou nulle. A cet égard des résultats négatifs ont été obtenus *in vitro* sur des bactéries et des cellules mammaliennes avec ou sans activation métabolique. L'épreuve de létalité dominante a également donné des résultats négatifs.

Des études de cancérogénicité à long terme (par voie orale et par inhalation) effectuées avec du CFC-11 et du CFC-12 sur des souris et des rats ont donné des résultats négatifs. Chez les rats, on a observé une réaction tumorigène au niveau de la cavité nasale après inhalation de CFC-113 mais cette réaction a été jugée douteuse. Les tumeurs présentaient une morphologie diversifiée et l'incidence n'était pas liée à la dose. Bien qu'on utilise des chlorofluorocarbures depuis plus de 50 ans, il n'existe qu'une seule étude de cohorte (539 travailleurs exposés). Elle n'a pas révélé d'augmentation de la mortalité globale ni de la mortalité par cancer.

Sur les huit chlorofluorocarbures examinés dans le présent document, seuls le CFC-11, le CFC-12 et le CFC-113 ont fait l'objet d'études de toxicité sur le développement publiées dans la littérature scientifique. Il n'existe de preuve d'embryotoxicité, de foetotoxicité ou de tératogénicité pour aucun de ces trois produits.

Résumé

8. Effets sur l'homme

Des études contrôlées sur des volontaires à qui l'on avait administré du CFC-11 et du CFC-12 n'ont pas révélé d'effets observables sur les paramètres hématologiques et biochimiques, le tracé électrocardiographique et électroencéphalographique, la fonction pulmonaire ni les paramètres neurologiques.

A forte concentration, les sujets ont ressenti des picotements et des bourdonnements d'oreille et ils ont éprouvé de l'appréhension. On a noté des modifications du tracé électroencéphalographique en même temps que des difficultés d'élocution et une réduction des performances dans les tests psychologiques. L'exposition à une concentration de 11% (545 g/m^3) de CFC-12 pendant 11 minutes[a] a provoqué une arythmie cardiaque notable suivie d'une baisse du niveau de conscience et d'une amnésie au bout de 10 minutes.

Après exposition à du CFC-12 à la concentration de 1% (50 g/m^3) pendant 150 minutes, on a observé une réduction de 7% dans les résultats d'épreuves psychomotrices, en revanche aucun effet n'a été noté à la dose de 0,1% (5 g/m^3).

Dans une étude au cours de laquelle 10 sujets ont été exposés à du CFC-11, du CFC-12, du CFC-14, deux mélanges de CFC-11 et de CFC-12 et ainsi qu'à un mélange de CFC-12 et de CFC-114 (concentrations dans l'air inspiré comprises entre 16 et 150 gr/m^3) pendant 15, 45 ou 60 secondes, on a observé une réduction sensible de la capacité ventilatoire pulmonaire (FEF50, FEF25) dans chaque cas, ainsi qu'une bradycardie, une irrégularité accrue du rythme cardiaque et un bloc auriculo-ventriculaire.

Après exposition à des concentrations de 0,15% (12 g/m^3), 0,25% (19 g/m^3), 0,35% (27 g/m^3) et 0,45% (35 g/m^3) de CFC-113 pendant 165 minutes, on a procédé à l'évaluation des performances psychomotrices. Aucun effet n'a été constaté à la concentration la plus faible; en revanche, les sujets éprouvaient une difficulté à se

---

[a] Tout au long de la présente monographie, les pourcentages de chlorofluorocarbures dans l'air sont exprimés au moyen du quotient du volume de chlorofluorocarbures par le volume d'air.

concentrer et les résultats des tests ont été un peu moins bons à partir de 0,35% (27 g/m^3).

D'après des études de portée limitée, il semblerait que les sujets ayant déjà eu des réactions allergiques cutanées aux déodorants en aérosol contenant du CFC-11 ou du CFC-12, puissent être sensibilisés à des applications cutanées de certains chlorofluorocarbures. Chez cinq non-fumeurs, l'exposition à du CFC-11 n'a pas perturbé la fonction muco-ciliaire trachéenne.

D'après deux études, il semblerait qu'une exposition professionnelle normale au CFC-113 ne comporte aucun risque sérieux pour la santé. Aucun effet indésirable n'a été noté à des taux d'exposition professionnelle atteignant 0,47 (36,7 g/m^3), le taux moyen d'exposition étant de 0,07% (5,4 g/m^3).

Plusieurs études ont fait état d'une diminution importante de la capacité ventilatoire chez des coiffeurs qui utilisaient des bombes aérosol contenant des chlorofluorocarbures. On a signalé des effets neurologiques après exposition professionnelle à des chlorofluorocarbures. C'est ainsi qu'on a décrit un cas de neuropathie chez un employé d'une blanchisserie exposé à du tétrachloroéthylène et à des concentrations indéterminées de CFC-113 pendant six ans.

Des cas d'exposition non-professionnelle accidentelle ou abusive consécutive à l'inhalation d'aérosols sont également attestés, les principaux symptômes étant une dépression du système nerveux central et des effets cardiovasculaires. Ces réactions indésirables, susceptibles parfois de conduire à la mort, sont attribuées à une arythmie cardiaque éventuellement aggravée par une élévation des catécholamines imputables au stress ou par une hypercapnie modérée.

L'accroissement du rayonnement UV-B devrait entraîner des effets essentiellement nocifs pour la santé humaine mais notre connaissance de ces divers effets est très variable. Pratiquement personne ne conteste que l'incidence des cancers cutanés non-mélanomateux augmenterait. Des projections basées sur des données récentes montrent que l'incidence de ces cancers augmenterait de 3% pour une diminution de 1% de la couche d'ozone. Il s'ensuit qu'une diminution de 5% de la couche d'ozone entraînerait chaque année dans le monde, au bout de quelques décennies, 200 000 cas supplémentaires de cancers cutanés non mélanomateux.

Le rayonnement UV-B joue également un rôle dans la formation des mélanomes cutanés qui sont encore plus dangereux. Toutefois on n'est pas encore en mesure de dégager des relations dose-réponse précises.

Le rayonnement UV-B peut influer de diverses manières sur le système immunitaire. Bien que, faute de connaissances suffisantes, on ne soit pas encore en mesure de prévoir quelles seraient exactement les conséquences d'une réduction de la couche d'ozone pour la santé humaine, il est probable qu'il s'en suivrait une augmentation de l'incidence des maladies infectieuses. Au niveau de l'oeil, l'effet le plus important serait un accroissement de l'incidence des cataractes, une opacification permanente du cristallin qui, même au niveau actuel de rayonnement UV-B, entraîne chez un grand nombre de personnes une réduction de l'acuité visuelle, voire la cécité.

En outre, l'accroissement du rayonnement ultra-violet favoriserait la formation du smog photochimique, ce qui aggraverait encore les problèmes de santé qui lui sont liés dans les agglomérations urbaines et les zones industrielles.

9. Evaluation des risques pour la santé humaine

Les effets directs les plus importants pour l'homme d'une exposition à des chlorofluorocarbures proviennent des concentrations excessives qui résultent d'accidents survenus dans l'industrie ou d'une utilisation défectueuse ou abusive de ces produits comme solvants ou gaz propulseurs. La libération de chlorofluorocarbures dans l'environnement général lors du rejet de déchets ou au cours du transport et du stockage est une source croissante de préoccupation en raison des conséquences que ces émissions incontrôlées pourraient avoir pour l'avenir de l'humanité.

EVALUATION DES RISQUES POUR LA SANTE HUMAINE ET EFFETS SUR L'ENVIRONNEMENT

1. Evaluation des risques pour la santé humaine

1.1 Effets directs sur la santé résultant d'une exposition à des chlorofluorocarbures complètement halogénés

La cinétique et le métabolisme des chlorofluorocarbures se caractérisent par une résorption pulmonaire et une distribution rapides. Rien n'indique qu'il y ait la moindre accumulation. Les chlorofluorocarbures examinés dans la présente monographie ne subissent qu'une métabolisation négligeable, si tant est qu'elle se produise. En conséquence, les effets toxiques d'éventuels métabolites sont très improbables. Les chlorofluorocarbures ont une très faible toxicité aiguë comme le montrent les études effectuées sur diverses espèces animales et par diverses voies d'administration. Cette toxicité se caractérise par des effets sur le myocarde, sur le système respiratoire et occasionnellement, sur le foie. Ces effets correspondent à la symptomatologie observée lors d'intoxications aiguës chez l'homme.

Des expositions répétées conduisent à des symptômes cliniques comparables. On observe parfois des anomalies au niveau du foie et des reins. Chez l'homme, un sérieux abus de ces substances ou une exposition incontrôlée ou accidentelle d'origine professionnelle peuvent conduire à des symptômes neurologiques centraux, cardiovasculaires et respiratoires. Dans des conditions d'emploi où l'exposition, de brève durée, ne dépasse pas 1000 ppm, il ne devrait pas se produire d'effets indésirables sur la santé.

D'après les études effectuées sur l'animal, on estime qu'il n'existe pas de risque cancérogène pour l'homme. Cette conclusion est corroborée par le fait que les chlorofluorocarbures étudiés dans la présente monographie sont dépourvus de génotoxicité ainsi que le montre l'étude des différents paramètres de la mutagénèse et de la transformation cellulaire. Lors d'une étude de cohorte limitée à 539 travailleurs exposés, on n'a pas constaté d'accrois-

Evaluation

sement de la mortalité ni de la proportion des tumeurs. Des études portant sur l'influence de ces produits sur la fonction de reproduction (fécondité, embryotoxicité, feototoxicité, tératogénicité) et sur les effets exercés au niveau du développement en général, ont donné des résultats systématiquement négatifs. On n'a pas connaissance d'effets sur la reproduction humaine, notamment pendant la vie intra-utérine ou au cours du développement post-natal.

On a observé dans l'air ambiant de zones urbaines ou suburbaines des concentrations moyennes de l'ordre de 3,4 $\mu g/m^3$ pour ce qui concerne le CFC-11 et de 6 $\mu g/m^3$ pour le CFC-12. Dans les régions rurales ou écartées, les valeurs correspondantes se situaient pour le CFC-11 à 1,0 $\mu g/m^3$ et pour le CFC-112 à 1,6 $\mu g/m^3$.

Ces taux d'exposition sont considérés comme négligeables par rapport aux concentrations de 25 000 à 50 000 $\mu g/m^3$ (équivalents à 5000 à 10 000 ppm) nécessaires pour faire apparaître les premiers signes d'anomalies fonctionnelles ou morphologiques chez les animaux de laboratoire.

1.2 Effets sur la santé découlant d'une réduction de l'ozone stratosphérique sous l'action des chlorofluorocarbures

Au cours de la dernière décennie, une inquiétude croissante s'est manifestée quant aux conséquences d'une réduction de la couche d'ozone dans la haute atmosphère, qui entraînerait du même coup un accroissement du rayonnement UV-B à la surface du globe. Selon le scénario adopté pour les émissions de chlorofluorocarbures et d'autres gaz en traces, les modèles utilisés permettent de prévoir que d'ici une cinquantaine d'années, la réduction de la couche d'ozone sera de 1 à 10%.

Parmi les effets possibles sur la santé humaine, on a très largement étudié l'induction de cancers cutanés non mélanomateux, tant du point de vue épidémiologique chez l'homme qu'au laboratoire sur l'animal. On admet en général que l'incidence des cancers non mélanomateux augmentera par suite de la réduction de la couche d'ozone. Selon une estimation fondée sur des données récentes, on prévoit qu'une réduction de l'ozone atmosphérique de 1% conduirait à une augmentation de 3% de l'incidence des cancers cutanés non mélanomateux. Une réduction de 5% de la couche d'ozone augmenterait l'incidence de ces cancers de 16%. Cette dernière valeur signifie qu'au niveau mondial, on enregistrerait plus de 200 000 nouveaux cas de cancer de ce type chaque année, principalement chez les individus à peau claire.

On a de plus en plus de raisons de penser que le rayonnement UV-B joue également un rôle dans l'induction et le développement des mélanomes, un cancer cutané beaucoup plus grave. Toutefois les incertitudes qui subsistent quant à la relation dose-effet rendent très difficiles les prévisions quantitatives. Il faut en tout cas prendre en compte la possibilité d'une augmentation de la fréquence des mélanomes.

Chez l'animal d'expérience, le rayonnement UV-B produit divers types d'immunosuppression. Celle-ci se traduit: (a) par une moindre résistance aux tumeurs implantées induites par les UV-B et par un plus grand développement de ces tumeurs chez les souris; (b) par la suppression de la sensibilisation par les allergènes de contact et (c) par la suppression de la réaction aux allergènes chez les animaux sensibilisés. On constate également une perturbation de la réponse immunitaire contre certains agents infectieux, qu'on a pu mettre en évidence dans le cas d'*Herpes simplex* et de *Leishmania* sp. On peut en déduire qu'une immunosuppression du même type pourrait se produire chez l'homme par suite d'exposition au rayonnement UV-B. Les cellules langerhansiennes cutanées qui présentent l'antigène sont endommagées et il s'ensuit une dépression des réponses allergiques. Bien qu'on ait encore beaucoup à apprendre sur ces phénomènes, il ne faut pas négliger la possibilité d'effets immunosuppresseurs et, par voie de conséquence, une augmentation dans l'incidence de certaines maladies infectieuses par suite d'une diminution de l'ozone stratosphérique.

On a des raisons de penser que le rayonnement UV-B favorise la formation de cataractes, une cause importante de cécité, en particulier dans les régions peu médicalisées.

Evaluation

2. Effets sur l'environnement

A part la théorie selon laquelle les chlorofluorocarbures contribuent à l'effet de serre, on ne dispose d'aucune preuve d'effets écologiques directs imputables aux chlorofluorocarbures examinés dans la présente monographie. Les études relatives aux effets du rayonnement UV-B sur les végétaux sont essentiellement consacrées aux plantes cultivées et ont généralement été menées sous des latitudes tempérées. Ces régions ne représentent qu'une faible part des grands écosystèmes de la planète. Bien qu'en raison de la complexité des études expérimentales, il subsiste un grand nombre d'incertitudes à cet égard, les données qui sont d'ores et déjà disponibles montrent que le rendement des récoltes aurait à souffrir d'un accroissement du rayonnement UV-B solaire. Sur plus de 200 espèces et variétés dont on a étudié la tolérance au rayonnement ultra-violet, les deux tiers environ s'y sont révélés sensibles. Parmi les plantes les plus sensibles figurent les pois, les haricots, les melons, la moutarde et le chou. Les membres de la famille des graminées sont généralement moins sensibles.

On peut observer expérimentalement qu'il existe une certaine tolérance au rayonnement UV-B dans le patrimoine génétique. Cette conclusion est tirée du fait qu'il existe des variations importantes dans la sensibilité au rayonnement UV parmi les diverses variétés de plantes cultivées. Les fondements génétiques de cette sensibilité restent à élucider.

On a étudié l'effet d'un accroissement du rayonnement UV-B sur la qualité des récoltes. En exposant diverses variétés de soja à un rayonnement UV qui résulterait d'une réduction de 25% de la couche d'ozone, on a constaté que la teneur en protéines et en huile des graines de soja diminuait dans une proportion pouvant atteindre 10%.

Les études relatives aux effets du rayonnement UV-B sur la production forestière restent limitées. Les résultats obtenus ne concernent que les jeunes plants et correspondent à des niveaux d'exposition qui résulteraient d'une réduction de 40% de la couche d'ozone. Ces travaux font état d'une réduction de la croissance et de la photosynthèse après exposition de plants de *Pinus taeda*. L'expérience montre qu'un accroissement du rayonnement UV-B peut provoquer une modification dans la structure des populations arborales.

On a montré que le rayonnement UV-B pouvait affecter les constituants végétaux et animaux des écosystèmes marins. Ces effets se traduisent par une réduction de la fécondité, de la croissance, de la survie et d'autres paramètres.

3. Conclusions

Les données toxicologiques sur les chlorofluorocarbures complètement halogénés examinées dans la présente monographie montrent que ces produits n'ont qu'une faible toxicité aiguë et chronique et qu'ils sont dépourvus d'activité mutagène ou cancérogène. Les risques pour la santé humaine sont limités aux cas d'exposition occasionnelle à de fortes concentrations susceptibles de se produire lors de la manipulation de ces produits. En revanche, les effets indirects résultant de l'accumulation de ces substances dans la stratosphère pourraient entraîner des effets non négligeables sur la santé humaine dus principalement à l'accroissement du rayonnement UV-B résultant de la réduction de la couche d'ozone stratosphérique. L'accroissement prévisible de l'incidence des cancers cutanés non mélanomateux, la possibilité d'une augmentation des mélanomes ainsi que les effets immunosuppresseurs et ophtalmologiques, sont autant d'éléments qui incitent à conclure à la nécessité d'une coopération internationale immédiate et efficace pour éviter que la couche d'ozone ne se réduise davantage.

RECOMMANDATIONS

1. La base de données toxicologiques relative à certains chlorofluorocarbures, en particulier ceux qui contiennent de l'hydrogène, est insuffisante pour permettre une évaluation quantitative du risque. Il faudrait obtenir davantage de renseignements sur la toxicité chronique, la cancérogénicité, les effets tératogènes et les effets sur la fonction de reproduction de ces composés, en particulier par suite d'expositions par inhalation.

2. On trouvera résumée au Tableau 16 une évaluation des effets exercés par l'accroissement du rayonnement UV-B.

Tableau 16. Effets potentiels d'un accroissement du rayonnement UV-B résultant d'une diminution de la couche d'ozone stratosphérique[a]

Effets	Niveau des connaissances	Impact planétaire potentiel
Cancers cutanés	Moyen à élevé	Moyen
Système immunitaire	Faible	Elevé
Cataracte	Moyen	Faible[b]
Flore[c]	Faible	Important
Faune et flore aquatiques[c]	Faible	Important
Impact climatologique[d]	Moyen	Moyen
Ozone ambiant	Moyen	Faible[e]

[a] Tiré, avec des modifications, de SAB-EC-87-025 *Review of EPA's Assessment of the Risks of Stratospheric Modification* par le Stratospheric Ozone Subcommittee, Science Advisory Board, US Environmental Protection Agency, Mars 1987.

[b] Une réflexion récente à propos de l'influence de la réduction de la couche d'ozone sur l'incidence de la cataracte conduit à considérer que l'impact pourrait en être plus grave (US EPA, Assessing the Risks of Trace Gases that can modify the Stratosphere, Chapitre 10, Decembre 1987).

[c] Voir section 6.

[d] Influence de la réduction de l'ozone stratosphérique elle-même et des gaz qui sont à l'origine de cette réduction sur le climat, y compris l'élévation du niveau des mers.

[e] L'impact pourrait être important dans certains secteurs urbains ou ruraux caractérisés par des problèmes de pollution par l'ozone au niveau du sol, à l'échelon régional ou local.

Il faut poursuivre les recherches dans les secteurs où les connaissances restent insuffisantes et où l'impact potentiel au niveau planétaire est important. Il s'agit de huit secteurs où une évaluation sera nécessaire dans l'avenir et qui sont particulièrement importants pour la compréhension des effets dus à la réduction de l'ozone atmosphérique et la conduite à tenir devant ces phénomènes; il faut donc:

- étudier les mécanismes de l'immunosuppression sur des modèles animaux et chez l'homme;
- déterminer quelles maladies infectieuses comportent un stade ou un processus susceptibles d'être aggravés par une exposition au rayonnement UV-B et mettre au point des modèles pour expliquer ces pathologies;
- étudier les effets d'une exposition au rayonnement UV-B sur l'incidence des maladies infectieuses et en particulier examiner dans quelle mesure elles dépendent de la longueur d'onde et établir des relations dose-réponse chez l'homme;
- déterminer l'impact de l'immunosuppression par le rayonnement UV-B sur l'efficacité des vaccins;
- clarifier le rôle des modifications immunologiques dans l'apparition des mélanomes et des cancers cutanés non mélanomateux sous l'action du rayonnement ultraviolet;
- déterminer le spectre d'action et les relations dose-effet dans le cas de l'induction de différents types de mélanomes par le rayonnement ultra-violet;
- mettre au point une meilleure définition des spectres d'action relatifs à l'induction des épithéliomas spinocellulaires et en particulier des épithéliomas cutanés baso-cellulaires par rayonnement ultra-violet;
- étudier la biologie et l'épidémiologie de la cataracte et développer des méthodes permettant de réduire le risque d'affections oculaires.

3. Certaines autorités recommandent encore l'utilisation du CFC-11 et du CFC-12 comme gaz propulseurs dans les bombes aérosol utilisées pour la désinsectisation des aéronefs. Il y a nécessité urgente à mettre au point un gaz propulseur d'un type nouveau, ininflammable, sûr, non irritant et qui se s'attaque pas à la couche d'ozone car les anciens produits sont d'ores et déjà interdits dans de nombreux pays.

4. Une coopération internationale efficace est nécessaire pour éviter que la couche d'ozone stratosphérique ne

Recommandations

se réduise davantage et pour cela, il faut réduire d'au moins 80 à 90% les émissions de chlorofluorocarbures qui détruisent l'ozone. Il faut en premier lieu trouver des substituts à ces produits, après quoi l'on devra imaginer des méthodes qui permettent d'éliminer dans de bonnes conditions les déchets de chlorofluorocarbures existants. Il est recommandé à tous les pays de prendre des dispositions pour réduire l'utilisation de chlorofluorocarbures ayant une forte tendance à détruire l'ozone stratosphérique.

RESUMEN

1. Identidad, propiedades físicas y químicas, y métodos analíticos

La presente monografía trata sólo de los clorofluorocarbonos (CFC) derivados de la sustitución completa de los átomos de hidrógeno del metano y el etano por átomos de flúor y cloro. Muchos de esos productos tienen importancia comercial y se sabe que algunos de ellos contribuyen a la disminución del ozono. Los productos examinados en el presente informe son los siguientes: triclorofluorometano (CFC-11), diclorodifluorometano (CFC-12), clorotrifluorometano (CFC-13), 1,2-difluoro-1,1,2,2-tetracloroetano (CFC-112), 1,1-difluoro-1,2,2,2-tetracloroetano (CFC-112a), 1,1,2-tricloro-1,2,2-trifluoroetano (CFC-113), 1,1,1-tricloro-2,2,2-trifluoroetano (CFC-113a), 1,2-dicloro-1,1,2,2-tetrafluoroetano (CFC-114), 1,1-dicloro-1,2,2,2-tetrafluoroetano (CFC-114a) y 1-cloro-1,1,2,2,2-pentafluoroetano (CFC-115). No se examinan los productos que no contienen cloro (como el CFC-134a y el CFC-116). Los productos que contienen hidrógeno (como el clorodifluorometano) se considerarán en un informe ulterior.

Los clorofluorocarbonos comerciales figuran entre los productos químicos orgánicos de mayor pureza disponibles. Se caracterizan habitualmente por una presión de vapor y una densidad elevadas y por valores bajos de viscosidad, tensión superficial, índice de refracción y solubilidad en agua. El grado de sustitución por flúor influye grandemente en las propiedades físicas y, en general, a medida que aumenta la sustitución por flúor, se eleva la presión de vapor y disminuyen el punto de ebullición, la densidad y la solubilidad en agua.

Los clorofluorocarbonos examinados en la presente monografía son razonablemente estables desde el punto de vista químico y, en ausencia de catalizadores metálicos, presentan bajas tasas de hidrólisis. Son muy resistentes al ataque por los agentes oxidantes convencionales en temperaturas inferiores a 200 °C. Por lo general, los clorofluorocarbonos presentan un elevado grado de estabilidad térmica y son extremadamente resistentes a casi todos los

Resumen

reactivos químicos. Sin embargo, reaccionan violentamente con los metales dotados de reactividad química.

Se dispone de varios métodos analíticos para la determinación de los clorofluorocarbonos en distintos medios. Comprenden la espectrofotometría, la cromatografía de gases con varios métodos de cuantificación y la espectrometría de masa. La mayor parte de los métodos emplean la cromatografía de gases con distintas técnicas de detección; los límites de detección suelen ser del orden de una parte por billón (ppb). Se han modificado los métodos de recogida de muestras para aumentar su selectividad y sensibilidad.

2. Fuentes de exposición humana y ambiental

Conforme a los conocimientos actuales, los clorofluorocarbonos examinados en la presente monografía no aparecen naturalmente en el medio ambiente, pero casi todos los clorofluorocarbonos, excepto los utilizados como productos intermedios químicos, pasan al medio ambiente. La producción mundial estimada de los clorofluorocarbonos con posibilidades importantes de reducción del ozono (CFC-11, CFC-12, CFC-113) en 1985 fue por lo menos de un millón de toneladas. La fabricación no se halla limitada a los principales países industriales y se realiza por lo menos en 16 países. Al aplicarse el Protocolo de Montreal probablemente se invertirá la actual tendencia al aumento en la fabricación de esos clorofluorocarbonos.

El método más importante de fabricación de los principales clorofluorocarbonos es el desplazamiento catalítico del cloro presente en los clorocarbonos por flúor mediante la reacción con fluoruro de hidrógeno anhidro. La mayor parte de la liberación al medio ambiente se produce durante la eliminación del equipo de desecho que contiene refrigerante y no en el curso de la fabricación, el almacenamiento ni la manipulación. La emisión de clorofluorocarbonos propulsantes ha disminuido como resultado de las restricciones legislativas impuestas a su uso en numerosos países, y la liberación de agentes de relleno es escasa. Dada la elevada presión de vapor de esos productos en las temperaturas ambientales, casi toda la cantidad liberada al medio ambiente se acumula en definitiva en la atmósfera. La emisión anual estimada en alrededor de un millón de toneladas en 1985 consistió principalmente en CFC-11 y CFC-12, y la liberación acumulativa de esos clorofluorocarbonos de 1931 a 1985 fue de 13,5 millones de toneladas aproximadamente.

La distribución mundial aproximada del uso de clorofluorocarbonos en 1985 fue la siguiente: refrigerantes, 15%; agentes de relleno de espuma, 35%; impulsores de aerosoles, 31%; varios, 7%, y sin designar, 12%. En los Estados Unidos de América, el empleo de impulsores de aerosoles fue muy inferior debido a las restricciones impuestas.

3. Transporte, distribución y transformación en el medio ambiente

Los clorofluorocarbonos comerciales persisten en el medio ambiente debido a su estabilidad química. Los tiempos medios de presencia en la atmósfera se calculan en 65, 110, 400, 90, 180 y 380 años para el CFC-11, el CFC-12, el CFC-13, el CFC-113, el CFC-114 y el CFC-115, respectivamente. Esos prolongados periodos de presencia aseguran la difusión a la estratosfera, en donde los clorofluorocarbonos reaccionarán con la capa de ozono por medio de los átomos de cloro liberados por un mecanismo fotoquímico. Además esos productos contribuirán al efecto de invernadero.

4. Niveles ambientales y exposición humana

Varios investigadores han señalado que los clorofluorocarbonos presentan una distribución mundial. Se han medido recientemente las variaciones latitudinales de las concentraciones de clorofluorocarbonos y se han hallado escasas diferencias en las concentraciones de CFC-11 y CFC-12 entre los hemisferios septentrional y meridional. Tampoco hay una variación notable en relación con la altitud hasta 6 km por encima de la superficie de la tierra. Las concentraciones medidas de clorofluorocarbonos en el aire de las zonas urbanas-suburbanas son superiores a las registradas en las zonas rurales-remotas debido a la contribución de las fuentes locales de emisión.

Las concentraciones atmosféricas de CFC-11 y CFC-12 aumentaron constantemente hasta 1985, año en el que los niveles combinados de esos dos productos en los Estados Unidos de América eran de 9120 ng/m^3 en las zonas urbanas-suburbanas, y de 2720 ng/m^3 en las zonas rurales-remotas para ambas sustancias. Partiendo de esos datos se ha calculado que la inhalación humana es de 182 y 54 mg/día en esos dos tipos de zonas.

Las concentraciones medias en la superficie oceánica de CFC-11 y CFC-12, registradas en tres emplazamientos distantes entre sí, eran del orden de O,2 ng/litro. Sin embargo, se midieron valores de 0,62 ng de CFC-11 por litro en el mar de Groenlandia en 1982 y hasta de 0,54 ng/litro en las aguas costeras del Japón. En esas mismas aguas se registró el valor máximo de CFC-12: 0,33 ng/litro. Se han medido niveles mucho más altos en las aguas dulces del lago Ontario: 249 mg de CFC-11 por litro y 572 ng de CFC-12 por litro. No se han detectado los clorofluorocarbonos en el agua de beber, pero se han hallado en la nieve y el agua de lluvia en Alaska, el lago Ontario y el río Niágara. Se ha detectado la presencia de CFC-11 en concentraciones de 0,1-5 µg/kg (ppb) (peso en seco) en distintos órganos de pescados y moluscos. Sin embargo, no se ha probado la presencia de clorofluorocarbonos en los alimentos tratados.

5. Cinética y metabolismo

Los clorofluorocarbonos pueden penetrar en el organismo humano por inhalación, ingestión o contacto cutáneo. La inhalación es la vía de entrada más corriente e importante, mientras que la espiración es la forma de eliminación del organismo más significativa. Los estudios controlados con voluntarios y animales de experimentación han proporcionado datos interesantes respecto a la exposición a distintos clorofluorocarbonos. Esos datos indican que los clorofluorocarbonos:

- pueden absorberse por la membrana alveolar, el tracto gastrointestinal o la piel;
- pasan con rapidez a la sangre, después de la inhalación;
- pasan a la sangre a una tasa decreciente al aumentar la concentración sanguínea;

- una vez presentes en la sangre, son absorbidos por distintos tejidos;
- alcanzan una concentración sanguínea estable si la exposición es suficientemente larga, indicando la existencia de un equilibrio entre el aire que contiene clorofluorocarbonos y la sangre;
- se absorben todavía por los tejidos orgánicos después de la estabilización inicial de la concentración sanguínea, y siguen penetrando en el organismo.

Los estudios efectuados en animales muestran que los clorofluorocarbonos se absorben con rapidez después de la inhalación y se distribuyen a través de la sangre en casi todos los tejidos del organismo. Las mayores concentraciones se encuentran habitualmente en los tejidos adiposos o que contienen lípidos. Sin embargo, los clorofluorocarbonos se hallan también en órganos bien irrigados, por ejemplo, el corazón, los pulmones, los riñones y la musculatura.

Los resultados de estudios metabólicos efectuados en el hombre y los animales han demostrado la resistencia de los clorofluorocarbonos a descomponerse o experimentar una transformación metabólica en los sistemas biológicos. Esos resultados permiten pensar que los clorofluorocarbonos se metabolizan en general en cuantía escasa o incluso nula después de la exposición.

Cualquiera que sea la vía de entrada, los clorofluorocarbonos se eliminan casi exclusivamente por las vías respiratorias en el aire espirado. No se ha señalado una recuperación significativa de clorofluorocarbonos o de sus metabolitos en los estudios que han tratado de identificar productos de transformación metabólica eliminados en la orina o las heces.

6. Efectos en el medio ambiente

Ciertos clorofluorocarbonos, en particular los CFC-11, 12, 113, 114 y 115, son extremadamente estables en las condiciones reinantes en la atmósfera baja. Los procesos fotolíticos que separan el cloro de los clorofluorocarbonos no se producen hasta que esos gases emigran al medio

Resumen

de radiación de alta energía de la estratosfera superior. Entonces los radicales cloro destruyen el ozono por catálisis. El ozono estratosférico absorbe la radiación ultravioleta solar (UV- B: 280-320 nm de longitud de onda), permitiendo que penetre hasta la superficie de la tierra sólo una cantidad reducida de radiación UV-B.

Los datos experimentales indican que un aumento de la irradiación UV-B en la superficie terrestre, resultante de la disminución del ozono, ejercería efectos nocivos en los biotas terrestres y acuáticos. Pese a las incertidumbres resultantes del carácter complejo de los experimentos prácticos, los datos actualmente disponibles permiten pensar que el rendimiento de las cosechas y la productividad de los bosques son vulnerables al aumento de la radiación de UV-B solar. Los datos existentes indican también que el incremento de la radiación UV-B modificará la distribución y abundancia de las plantas y cambiará la estructura del ecosistema.

Varios estudios de los ecosistemas marinos han demostrado que la radiación UV-B produce daños en las larvas de los peces y los peces de poca edad, las larvas de camarones y cangrejos, los copépodos y las plantas indispensables para la red alimentaria marina. Entre los efectos dañinos figuran el descenso de la fecundidad, el crecimiento y la supervivencia. Los datos experimentales indican que incluso pequeños aumentos de la exposición a la radiación UV-B ambiental pueden dar lugar a cambios notables del ecosistema.

7. Efectos en animales de experimentación y sistemas *in vitro*

Se ha estudiado ampliamente la toxicidad aguda por inhalación de clorofluorocarbonos. Los clorofluorocarbonos examinados en la presente monografía presentan una escasa toxicidad aguda por inhalación. La sintomatología de la intoxicación aguda comprende efectos en el sistema nervioso central (SNC), efectos secundarios en el sistema cardiovascular e irritación de las vías respiratorias. Los limitados datos disponibles sobre la toxicidad oral aguda de los clorofluorocarbonos muestran que es baja. Cuando se aplican en la piel en dosis altas, el CFC-112, el CFC-112a y el CFC-113 provocan distintos grados de irritación, pero ningún otro efecto notable.

Se han comunicado estudios de inhalación a corto plazo del CFC-11, el CFC-12, el CFC-112, el CFC-113, el CFC-114 y el CFC-115. Los resultados muestran una baja toxicidad y los efectos observados guardan relación principalmente con el SNC, las vías respiratorias y el hígado. Los estudios de toxicidad oral han confirmado la reducida toxicidad.

En un estudio de inhalación a largo plazo se expuso a ratas al CFC-113 al 0,2, 1 ó 2% (15,3, 76,6, ó 183 g/m^3), 6 horas por día, 5 días por semana, hasta 2 años. No se observaron efectos histopatológicos ni modificaciones de los valores de laboratorio clínico. La única observación que los autores consideraron relacionada con el tratamiento fue la disminución del aumento de peso corporal en los grupos expuestos a las dos dosis más altas.

Los datos disponibles muestran que los clorofluorocarbonos totalmente halogenados evaluados en la presente monografía tienen escaso o nulo potencial mutágeno o carcinógeno. Se han obtenido resultados negativos *in vitro* utilizando bacterias y células de mamífero, con o sin activación metabólica, en la prueba letal dominante.

Los estudios de cancerogenicidad a largo plazo (por vía oral y por inhalación) con CFC-11 y CFC-12, efectuados en ratas y ratones, dieron resultados negativos. Se observó una respuesta tumorígena en la cavidad nasal en ratas sometidas a la inhalación de CFC-113, pero esa reacción se consideró equívoca. Los tumores presentaron distintas morfologías y las incidencias no guardaban relación con la dosis. Aunque se utilizan los clorofluorocarbonos desde hace más de 50 años, sólo se dispone de un estudio de cohorte (539 trabajadores expuestos). No se observó ningún aumento de la mortalidad total ni de las defunciones por tumores.

Entre los ocho clorofluorocarbonos examinados en el presente documento, en las publicaciones científicas disponibles se han recogido estudios sobre efectos tóxicos en el desarrollo en los casos del CFC-11, el CFC-12 y el CFC-113. En ninguno de esos tres clorofluorocarbonos se han registrado indicios de embriotoxicidad, fetotoxicidad o teratogenicidad.

Resumen

8. Efectos en la especie humana

Los estudios controlados de voluntarios que utilizaron CFC-11 y CFC-12 no mostraron efectos observables en los parámetros hematológicos y químicos clínicos, el ECG, el EEG, la función pulmonar o la exploración neurológica.

En concentraciones altas, los sujetos experimentaron una sensación de picazón, zumbidos de oídos y aprensión. Se observaron modificaciones del EEG, alocución difícilmente inteligible y disminución de la habilidad en las pruebas psicológicas. La exposición a una concentración del 11%[a] (545 g/m^3) de CFC-12 durante 11 minutos provocó un grado importante de arritmia cardiaca, seguido de un descenso de la conciencia con amnesia al cabo de 10 minutos.

Tras la exposición al CFC-12 a una concentración del 1% (50 g/m^3) durante 150 minutos se observó un descenso del 7% en los índices de pruebas psicomotrices, pero ningún efecto con una concentración del 0,1% (5 g/m^3).

En un estudio en el que 10 sujetos estuvieron expuestos al CFC-11, el CFC-12 y el CFC-114, dos mezclas de CFC-11 y CFC-12, y una mezcla de CFC-12 y CFC-114 (concentraciones en el aire respirado comprendidas entre 16 y 150 g/m^3) durante 15, 45 ó 60 segundos, se registró en cada caso una reducción aguda importante de la capacidad pulmonar ventilatoria (FEF50, FEF25), así como bradicardia, aumento de la variabilidad del ritmo cardiaco y bloqueo auriculoventricular.

Se evaluó la habilidad psicomotriz utilizando el CFC-113 en concentraciones de 0,15% (12 g/m^3), 0,25% (19 g/m^3), 0,35% (27 g/m^3) ó 0,45% (35 g/m^3) durante 165 minutos. La concentración más baja careció de efecto, pero se produjeron dificultades para la concentración mental y cierto descenso en los resultados de las pruebas a partir de la dosis de 0,35% (27 g/m^3).

Los limitados estudios efectuados muestran que en las personas con antecedentes de reacción cutánea a los deso-

---

[a] En la totalidad de la presente monografía, los porcentajes de clorofluorocarbonos en el aire se expresan como el volumen de clorofluorocarbono dividido por el volumen de aire.

dorantes en pulverización que contienen CFC-11 o CFC-12, la aplicación cutánea de ciertos clorofluorocarbonos puede provocar una sensibilización. En cinco personas no fumadoras, la función mucociliar traqueal no se alteró por la exposición al CFC-11.

Dos estudios permiten pensar que la exposición profesional normal al CFC-113 no plantea un riesgo grave para la salud. No se observaron efectos adversos en niveles profesionales de hasta el 0,47% (36,7 g/m^3), con una concentración media del 0,07% (5,4 g/m^3).

En varios estudios se ha observado una disminución aguda importante de la capacidad pulmonar ventilatoria en los peluqueros que utilizan pulverizaciones para el pelo que contienen clorofluorocarbonos. Se han registrado casos de efectos neurológicos atribuidos a la exposición profesional a los clorofluorocarbonos. Se ha descrito un caso de neuropatía en un trabajador de lavandería expuesto al tetracloroeteno y a concentraciones indeterminadas de CFC-113 durante seis años.

Se ha notificado también la exposición no profesional y accidental o la inhalación por uso indebido de aerosoles, siendo los principales síntomas la depresión del SNC y las reacciones cardiovasculares. La arritmia cardiaca, agravada posiblemente por los niveles altos de catecolaminas provocados por el estrés o por la hipercapnia moderada, se considera la causa de esas respuestas adversas, que pueden conducir a la muerte.

Es de suponer que la mayor radiación de UV-B conducirá a efectos predominantemente adversos en la salud humana, pero el estado de conocimientos varía grandemente de un efecto a otro. Casi todos los autores admiten que aumentará la incidencia de los cánceres cutáneos distintos del melanoma. Las previsiones basadas en datos recientes muestran que esa incidencia se incrementará en un 3% por cada 1% de pérdida de ozono. Sobre esa base, una pérdida de ozono del 5% conduciría, al cabo de varios decenios, a la aparición cada año de más de 200 000 casos adicionales de cánceres cutáneos distintos del melanoma.

La radiación UV-B parece intervenir también en la formación de los melanomas cutáneos, tumores de mayor gravedad. Sin embargo, los conocimientos son insuficientes para establecer con precisión las relaciones dosis-respuesta.

El sistema inmunitario experimenta la influencia de la radiación UV-B de distintos modos. Aunque no se dispone de conocimientos suficientes para predecir las consecuencias de la disminución de ozono en la salud humana, se observará probablemente una mayor incidencia de enfermedades infecciosas.

El efecto más importante para el ojo humano será un aumento de la incidencia de la catarata, enturbiamiento permanente del cristalino del ojo que conduce, incluso con los actuales niveles de radiación UV-B, a alteración de la visión y ceguera en muchas personas.

Puede esperarse que el aumento de la radiación UV-B incremente el "smog" fotoquímico, lo que agravaría los problemas de salud conexos en las zonas urbanas e industrializadas.

9. Evaluación de los riesgos para la salud humana

Los efectos directos más importantes que provoca en el ser humano la exposición a los clorofluorocarbonos se deben a las concentraciones excesivas resultantes de accidentes industriales y del uso indebido o excesivo de esas sustancias como disolventes o gases impulsores. La liberación de clorofluorocarbonos en el medio ambiente mundial en el curso de la eliminación de desechos y del transporte y almacenamiento son motivo de creciente preocupación, debido a los posibles efectos que esas liberaciones incontroladas pueden ejercer en la salud futura de la humanidad.

EVALUACION DE LOS RIESGOS PARA LA SALUD HUMANA Y DE LOS EFECTOS EN EL MEDIO AMBIENTE

1. Evaluación de los riesgos para la salud humana

1.1 Efectos directos en la salud resultantes de la exposición a clorofluorocarbonos totalmente halogenados

La cinética y el metabolismo de los clorofluorocarbonos se caracterizan por la absorción y la distribución pulmonares rápidas. No hay indicio de ninguna acumulación. La transformación metabólica de los clorofluorocarbonos examinados en la presente monografía es despreciable, si es que realmente existe. Por consiguiente, los efectos tóxicos de los metabolitos son muy improbables. La toxicidad aguda de los clorofluorocarbonos es muy baja, como se demuestra en estudios efectuados en distintas especies animales con diferentes vías de administración. Se caracteriza por los efectos en el corazón, el sistema respiratorio y a veces el hígado. Esos efectos concuerdan con la sintomatología observada en intoxicaciones agudas en personas.

Tras la exposición repetida pueden observarse síntomas clínicos comparables. Se producen a veces alteraciones hepáticas y renales. En el hombre aparecen síntomas en el SNC, el sistema cardiovascular y el aparato respiratorio en los casos de uso indebido intenso y de exposición profesional incontrolada o accidental. En las condiciones de uso que suponen la exposición a corto plazo a una concentración de hasta 1000 ppm, no son de esperar afectos adversos en la salud.

La evaluación de los estudios efectuados en animales de experimentación no muestra que haya riesgo de cancerogénesis para el ser humano. Lo subraya el hecho de que los clorofluorocarbonos examinados en la presente monografía están desprovistos de genotoxicidad en distintos puntos finales mutagénicos y transformaciones celulares. En un estudio de cohorte limitado que comprendió 539 trabajadores expuestos, no se registró aumento de la mortalidad ni de la frecuencia de tumores. Los estudios sobre la influencia en la reproducción (fecundidad, embriotoxicidad, fetotoxicidad, teratología) y sobre los efectos generales en el desarrollo en animales de experimentación han sido constantemente negativos. No se han registrado efectos en la reproducción humana, incluido el desarrollo intrauterino y posnatal.

Se han medido concentraciones medias en el aire de las zonas urbanas-suburbanas de 3,4 $\mu g/m^3$ de CFC-11 y de 6 $\mu g/m^3$ de CFC-12. En las zonas rurales-remotas, los niveles correspondientes fueron de 1,0 $\mu g/m^3$ para el CFC-11 y de 1,6 $\mu g/m^3$ para el CFC-12.

Esos niveles de exposición se consideran despreciables en comparación con las concentraciones de 25 000 a 50 000 $\mu g/m^3$ ($\approx$5000 a 10 000 ppm) que causan signos iniciales de alteraciones funcionales o morfológicas en los animales de laboratorio.

1.2 Efectos en la salud previstos provocados por la reducción del ozono estratosférico causada por los clorofluorocarbonos

En el último decenio se ha producido una creciente preocupación por las consecuencias de la disminución del ozono en la atmósfera superior, con el aumento consiguiente de la radiación UV-B en la superficie de la tierra. Los cálculos en modelos predicen, para los próximos cinco años, una pérdida de ozono comprendida entre el 1% y el 10%, en función del supuesto utilizado para la liberación de los clorofluorocarbonos y de otros gases en oligoconcentraciones.

Entre los efectos en la salud humana se ha investigado ampliamente la inducción de cánceres cutáneos distintos al melanoma, tanto en epidemiología humana como en animales de experimentación. En general se ha aceptado la conclusión de que la incidencia de esos cánceres aumentará como resultado de la disminución del ozono. Una estimación basada en datos recientes prevé que una reducción del ozono atmosférico del 1% conduciría a un aumento de la incidencia de cánceres cutáneos distintos del melanoma del 3%. Una reducción del ozono del 5% llevaría a un incremento de la incidencia del 16%. Este último supondría un aumento mundial de más de 200 000 casos nuevos de cánceres cutáneos distintos al melanoma por año, sobre todo en las personas de piel clara.

Aumentan los indicios que permiten pensar que la radiación UV-B interviene también en la inducción y proliferación del melanoma cutáneo, tipo más grave de cáncer de la piel. Sin embargo, la incertidumbre existente, en particular en lo que respecta a la relación dosis-efecto, hace que las predicciones cuantitativas sean muy difíciles. Ahora bien, debe tomarse en cuenta la posibilidad de un aumento del melanoma cutáneo.

La radiación UV-B produce distintos tipos de supresión específica del sistema inmunitario en los animales de experimentación. Se observa una disminución de la resistencia a los tumores implantados inducida por la radiación UV-B y un mayor crecimiento de tales tumores en los ratones; además se suprime la sensibilización por los alergenos de contacto y la respuesta a los alergenos en los animales sensibilizados. También se altera la respuesta inmunitaria frente a ciertos agentes infecciosos, como se ha demostrado en los casos del virus del herpes simple y de las leishmanias. Existen indicios de que la radiación UV-B puede producir en el hombre una supresión análoga de la respuesta inmunitaria. En la piel, las células de Langerhans de presentación de antígenos quedan lesionadas y disminuyen las respuestas alérgicas. Aunque todavía queda mucho por aprender en ulteriores investigaciones, no deben ignorarse los posibles efectos de supresión inmunitaria y el aumento consiguiente de la incidencia de ciertas enfermedades infecciosas que podrían resultar de la disminución del ozono estratosférico.

Ciertos datos muestran que la radiación UV-B aumenta la formación de la catarata, importante causa de ceguera, en particular en las zonas que poseen limitados servicios médicos.

2. Efectos en el medio ambiente

Aparte de la teoría de que los clorofluorocarbonos contribuyen al efecto de "invernadero", no se dispone de datos que señalen otros efectos ecológicos directos provocados por los clorofluorocarbonos examinados en la presente monografía.

Los estudios sobre los efectos de la radiación UV-B en las plantas se han concentrado en los cultivos y se han realizado en general en latitudes templadas. Estas representan sólo una pequeña porción de los principales ecosistemas del mundo. Aunque existen numerosas incertidumbres resultantes del carácter complejo de los experimentos, los datos actualmente disponibles permiten pensar que los cultivos son posiblemente vulnerables a mayores niveles de radiación UV-B solar. Entre más de 200 especies y cultivares examinados respecto a la tolerancia a la radiación ultravioleta, alrededor del 65% resultaron sensibles. Entre los grupos de plantas más sensibles figuraban los cultivos de guisantes, judías, melones, mostaza y coles. Los miembros del sistema de la hierba son en general menos sensibles.

Los datos experimentales muestran que el patrimonio genético presenta cierto grado de tolerancia a la radiación UV-B. Se basan en el alto grado de variación en la sensibilidad a la radiación UV observada en los cultivares. Todavía tiene que determinarse la base genética de la sensibilidad.

Se ha estudiado el efecto de los mayores niveles de radiación UV-B sobre la calidad de las cosechas. El contenido en proteínas y aceite de cultivares seleccionados de semilla de soja se redujo hasta en el 10% al exponer las plantas a niveles de radiación UV que simulaban una pérdida de ozono del 25%.

Se han realizado limitados estudios sobre los efectos de la radiación UV-B en la productividad forestal. Sólo se dispone de resultados relativos a los plantones y corresponden a niveles de exposición equivalentes a una reducción del ozono del 40%. Esos estudios muestran una disminución del crecimiento y de la fotosíntesis después de la exposición de plantones de *Pinus taeda*. Ciertos datos experimentales muestran que el aumento de los niveles de radiación UV-B puede producir cambios en la estructura del conjunto del bosque.

Se ha observado que la exposición a la radiación UV-B afecta a los componentes vegetales y animales de los ecosistemas marinos. Entre los efectos figuran los descensos de la fecundidad, el crecimiento, la supervivencia y otros parámetros.

3. Conclusiones

Los datos toxicológicos disponibles sobre los clorofluorocarbonos totalmente halogenados examinados en la presente monografía muestran que la toxicidad aguda y crónica es baja y no indican que tengan capacidad mutágena ni cancerígena. Los riesgos para la salud humana están principalmente limitados a exposiciones altas ocasionales, que pueden producirse al manipular esas sustancias. Por el contrario, los efectos indirectos que aparecen en el hombre por la acumulación de tales productos en la estratosfera pueden conducir a alteraciones notables de la salud humana, producidas sobre todo por la disminución del ozono estratosférico, que da lugar a un aumento de los efectos de la radiación UV-B. El aumento previsto en la incidencia de los cánceres cutáneos distintos del melanoma, el posible incremento del melanoma, y los efectos inmunotóxicos y oculares conducen a la conclusión de que es necesaria una cooperación internacional inmediata y eficaz para reducir toda nueva pérdida del ozono estratosférico.

RECOMENDACIONES

1. La base de datos sobre la toxicidad de algunos clorofluorocarbonos, en particular de los que contienen hidrógeno, es insuficiente para efectuar evaluaciones cuantitativas del riesgo. Se necesita información adicional sobre la toxicidad crónica, la cancerogenicidad y la teratogenicidad/efectos reproductores de los productos, en particular en el caso de la exposición por inhalación.

2. En el cuadro 16 se resume la evaluación de los efectos del aumento de la radiación UV-B.

Cuadro 16. Posibles efectos del aumento de la radiación UV-B resultantes del descenso del ozono estratosférico[a]

Efectos	Conocimientos disponibles	Posible impacto mundial
Cáncer cutáneo	Moderados a altos	Moderado
Sistema inmunitario	Escasos	Alto
Catarata	Moderados	Bajo[b]
Vida vegetal[c]	Escasos	Alto
Vida acuática[c]	Escasos	Alto
Repercusiones climáticas[d]	Moderados	Moderado
Ozono ambiental	Moderados	Bajo[e]

[a] Modificado de SAB-EC-87-025 *Review of EPA's Assessment of the Risk of Stratospheric Modification*, Stratospheric Ozone Subcommittee, Science Advisory Board, US Environmental Protection Agency, marzo de 1987.

[b] Un estudio más reciente sobre la influencia de la pérdida de ozono en la incidencia de la catarata permite pensar que puede ser más grave (US EPA, Assessing the Risks of Trace Gases that can modify the Stratosphere, Capítulo 10, diciembre de 1987).

[c] Véase la sección 6.

[d] Contribución a los cambios climáticos, incluido el aumento del nivel del mar, de la propia pérdida del ozono estratosférico y de los gases que causan esa pérdida.

[e] El impacto puede ser alto en determinadas zonas urbanas o rurales en las que son habituales los problemas de contaminación atmosférica por el ozono en el nivel superficial en escala local o regional.

Se necesitan más investigaciones en los sectores en los que faltan conocimientos y en los que el posible impacto mundial es elevado. Incluyen ocho sectores concretos de futuras investigaciones y evaluaciones que tienen particular importancia para conocer y afrontar los efectos en la salud humana de la pérdida del ozono estratosférico:

- investigar los mecanismos de la inmunosupresión en modelos animales y en el hombre;
- identificar las enfermedades infecciosas que comprenden una fase o proceso que puede empeorar por la exposición a la radiación UV-B y elaborar modelos para explicar esas enfermedades;
- investigar la dependencia respecto a la longitud de onda y obtener datos de dosis-respuesta para el hombre relativos a los efectos de la exposición a la radiación UV-B sobre la incidencia de las enfermedades infecciosas;
- determinar el efecto de la inmunosupresión por la radiación UV-B sobre la eficacia de la vacunación;
- aclarar la función de los cambios inmunológicos en la inducción de melanomas y de cánceres cutáneos distintos al melanoma por la radiación UV;
- determinar el espectro de acción y las relaciones dosis-efecto para la inducción de distintos tipos de melanoma por la radiación UV;
- establecer una definición mejor del espectro de mecanismos para la inducción del carcinoma escamocelular, y en particular del carcinoma basocelular, por la radiación UV; e
- investigar la biología y epidemiología de la catarata, y los métodos para reducir los riesgos de enfermedades oculares.

3. Algunos organismos recomiendan todavía el empleo del CFC-11 y del CFC-12 como propulsores para la desinfección de aeronaves por pulverizaciones en aerosol. Se necesitan urgentemente para ese uso nuevos propulsores que no reduzcan el ozono, ininflamables, inocuos y no irritantes, puesto que los antiguos gases propulsores están ya prohibidos en muchos países.

4. Esnecesaria la cooperación internacional efectiva para reducir la futura pérdida del ozono estratosférico, lo que exige reducciones del 80%-90% por lo menos en la emisión de clorofluorocarbonos reductores del ozono.

Recomendaciones

La primera prioridad consiste en hallar productos de sustitución y la segunda en elaborar procedimientos de evacuación apropiados para los actuales clorofluorocarbonos de desecho. Se recomienda que todos los países adopten medidas para reducir el empleo de los clorofluorocarbonos con altas posibilidades de reducción del ozono estratosférico.

www.ingramcontent.com/pod-product-compliance
Ingram Content Group UK Ltd.
Pitfield, Milton Keynes, MK11 3LW, UK
UKHW021311180426
11947UKWH00015B/1160